Woven Fabric Engineering

Woven Fabric Engineering

Editor

Abdulkerim Macar

scitus
academics

Woven Fabric Engineering
Edited by **Abdulkerim Macar**

ISBN: 978-1-68117-674-1
Library of Congress Control Number: 2016946540

616, Corporate Way, Suite 2, 4766,
Valley Cottage, NY 10989

www.scitusacademics.com

Notice

Reasonable efforts have been made to publish reliable data and views articulated in the chapters are those of the individual contributors, and not necessarily those of the editors or publishers. Editors or publishers are not responsible for the accuracy of the information in the published chapters or consequences of their use. The publisher believes no responsibility for any damage or grievance to the persons or property arising out of the use of any materials, instructions, methods or thoughts in the book. The editors and the publisher have attempted to trace the copyright holders of all material reproduced in this publication and apologize to copyright holders if permission has not been obtained. If any copyright holder has not been acknowledged, please write to us so we may rectify.

Printed in United States of America on Acid Free Paper

Preface

Woven fabric is a fabric made from tissue. It is created on a loom and made of a lot of weft and warp threads. Woven fabrics are classified as weaving or possibly structure based on the way weft and warps are crossed. The three basic tissues, of which other variations are, are plain, twisting and satin. In plain tissue, commonly known as calico, tabby, taffeta or possibly homespun tissues, the weft passes through alternate warp threads, requiring only two harnesses. The simple construction is suitable for cheap fabrics, heavy threads and printed designs. Variations are produced by the use of yarn groups, such as in basket tissue and monk's cloth, or by alternating fine and coarse yarns to make corded and ribbed fabrics, such as the warp ribbed Bedford cord, piqu, and dimity, as well as the weft ribbed poplin, rep, and grosgrain. The next primary weave, twill, shows a diagonal design that causes weft threads to interlace two to four warp threads, move a step to the right or maybe to the left of each pick and, for example, herringbone, capable of variations and corkscrew designs. Noted for the company, twill fabrics include close tissue, gabardine, drill, serge and denim. Satin fabric has overflowing or floating warp threads on the surface that reflect light and give a distinctive luster. The weave is called sateen when the uncrossed threads are in the weft.

This book emphasizes revolutionary concepts and cutting-edge findings in the arena of woven textile engineering, mainly for specialized applications, as well as in the field of fabric engineering and to address some issues related to the use of tissue in composite materials.

Table of Contents

Chapter 1

Smart Woven Fabrics in Renewable Energy Generation

Derman Vatansever[1], Elias Siores[1], Ravi L. Hadimani [1] and Tahir Shah[1]

[1] University of Bolton, Institute for Materials Research and Innovation, United Kingdom

1. INTRODUCTION

Initially, the first purpose of fabric making was only for covering the body and sheltering. However, with a growing population and ever improving advanced technologies, today's fabrics are mostly used for fashion and performance thus enhancing the standard of people's everyday life and enjoyment.

Most of the technologies which increase the standard of living also increase carbon emission and adversely affect human life indirectly. Warming buildings, using cars, provide hot water, cooking food etc. need energy generated by using coal, gas, fuel or electricity. Burning gas and fuel accelerate the threat of nature and eventually contribute to global warming. Even electricity generation causes carbon emission unless it is generated by using renewable energy sources. Interest in providing renewable usable electrical power from the environment has grown, particularly in the elimination of battery usage, because of their sizeable dimensions, weight and limited lifetime.

Since global warming is being considered as the biggest danger for the nature, many scientists and researchers have brought a new breath to their researches. As almost all areas of renewable science and technology, researchers are now working in the field of textile fabrics capable of generating green electricity.

Undoubtedly, weaving is the oldest fabric making method which has been a part of human life for protection from nature's elements and hazards. It is now possible to produce smart woven fabrics by combining the oldest fabric making method with smart fibre material technologies. The chapter named "Smart Woven Fabrics in Renewable Energy Generation "contains a brief introduction to smart materials, focusing on piezoelectricity and polymer based piezoelectric fibre production. The rest of the chapter explains how to produce smart woven structures by integrating smart fibres into the fabric during weaving process and examples for possible applications for energy regeneration from nature's elements are given.

2. BRIEF INTRODUCTION TO WEAVING AND LOOMS

2.1. Vertical and Horizontal Looms

The first loom consisted of only a branch of a tree that is parallel to the ground. In this simple design, warp threads were directly fastened to the branch of a tree and held parallel to each other under tension caused by tied stones at the other end of the warps. The weft threads work from right to left and left to right by passing through hanging warps until the designed fabric is composed. Later, a simple framework was made of tree branches to form a vertical loom to use for indoor as well as outdoor.

Figure 1 (a and b) show examples for early vertical loom and horizontal looms and Figure 1 shows a laboratory scale hand loom (c) and automated loom (d). Warp threads are attached to heddles and held parallel to each other on the loom and weft thread is held by a shuttle. The heddles keep warp threads apart so that the shuttle could pass through easily and cary the thread from left to right and right to left so that the weft will work over and under warp threads to form a fabric. Movements of the heddles on hand looms are controlled by pedals beneath the loom.

Current generation looms (Fig. 1d) are fully automated. Movements of the heddles are automatically controlled according to the fabric design sent via a computing system. The important parameters, such as tension, speed, temperature etc., are measured and/or controlled by the controlling systems. Thus, any errors that occur during the fabric production are immediately detected.

(a) (b)

Figure 1. a) A vertical warp-weighted loom used in Iceland, reconstructed and set up before 1914, National Museum of Iceland-Photograph by Gilsli Gestsson (Broudy, 1979),(b) a horizontal ground loom used in Northen Cameroon-Photograpgh by René Gardi (Broudy, 1979).

2.2. Main Weaving Constructions

Weaving is one of the traditional fabric making methods. There are two sets of threads, warp threads and weft threads which form a fabric by being interlaced row by row. Weaving has three main construction techniques; plain weaving, twill weaving and satin weaving. Any other techniques developed are variations of these main techniques.

Plain weave is created by interlacing the weft across the warp threads and for this at least 2 heddles are needed. Weft thread goes under a warp thread and then over the next one so that the equal amount of weft and warp is seen on both surfaces of the woven fabric.

(c)

(d)

Figure 1. c) An example of laboratory scale hand looms and (d) an example of automated looms (wikipedia, 2011).

Twill weave is created by interlacing two or more weft threads over and under one or more warp threads so that at least three heedles are needed to make a twill woven fabric. In twill wovens, parallel diagonal ribs are formed from left to right or from right to left that depends on the formation, woven fabric is characterised and named as "S" or "Z" twill, respectively. If a twill woven has more warps than the wefts on the fabric face is known as warp faced twill and if it has more wefts on the fabric face, it is known as weft faced twill.

Satin weave is created by floating four or more yarns before a single interlacing occurs and at least 5 heddles are needed to make a satin woven fabric. The warp or weft threads pass across many threads in such a way, like 4/1, 5/1, 7/1 etc., that the face of the woven fabric is mostly covered with either warp or weft threads and known as warp satin or weft satin, respectively.

Depending on the construction of the weave and characteristics of the used warp and weft threads, fabric properties may vary. Count of the used fibres/yarns determines the fabric density which affects the weight, feel, appearance, thickness etc. Variable properties are many and can be found in the literature (Gioello, 1982). They include:

- Weight
- Hand and feel
- Drapeability
- Appearance
- Covering power
- Surface texture
- Body fit
- Thickness
- Lustre
- Strength
- Flexibility / pliability
- Resiliency
- Warmth
- Affinity to dyestuff

Fabrics created by using plain weave technique are tough compared to twill and satin weave. The fabrics made by twill weaving technique are more pliable and drapable than plain weave fabrics but not as pliable as satin weave. However, strength of plain woven fabrics is higher than both twill and satin woven fabrics.

3. SMART MATERIALS AND PIEZOELECTRICITY

A smart material is one that shows extraordinary response when subjected to a stimulus. Piezoelectric materials are considered as smart materials because of their ability to generate electricity against the stimulus which is mechanical strain or vibration (Fig.2a). This property of piezoelectric materials is known as direct piezoelectric effect. The reverse effect is also possible in that this material undergoes a slight deformation in shape when a small electrical field is applied (Fig.2b).

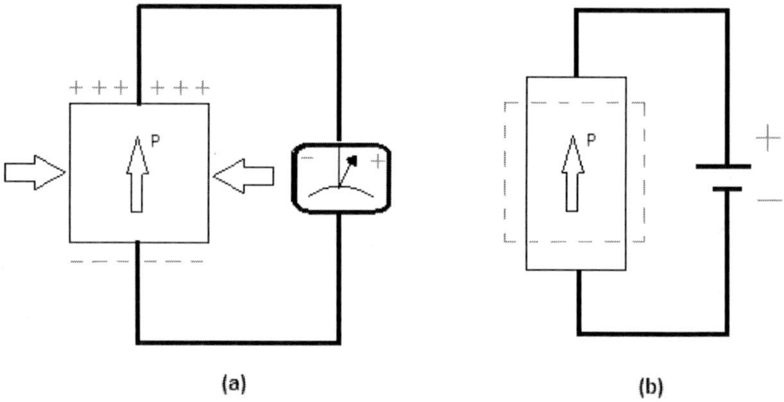

Figure 2. a) Direct piezoelectric effect; mechanical energy is converted to form electrical energy, (b) Converse piezoelectric effect; electrical energy causes deformation in the shape of piezoelectric material.

Direct piezoelectric effect of the materials is also known as "generator effect" or "sensor effec". The terms "motor effect "and" ctuator effect" are also used to mean converse piezoelectric effect of the piezoelectric material. It should also be pointed out that the piezoelectric effect can be induced when heat or cooling is involved, in which case this phenomenon is termed thermoelectric or pyroelectric effect.

Direct piezoelectric effect is mainly used for energy harvesting. The term "Energy Harvesting" is used to describe the process of extracting energy from the environment and the extracted energy is converted and stored in the form of electrical energy.

Although energy harvesting technologies have been known for many years, increasing concern about global warming has led to intensive research for alternative energy sources including piezoelectrics. With an increasing concern about global warming, piezoelectricity has gained a significant importance and intensive research and development efforts are being made for extracting energy from the environment [Umeda et al., 1997; Sodano et al., 2004 ; Mateu & Moll, 2005].

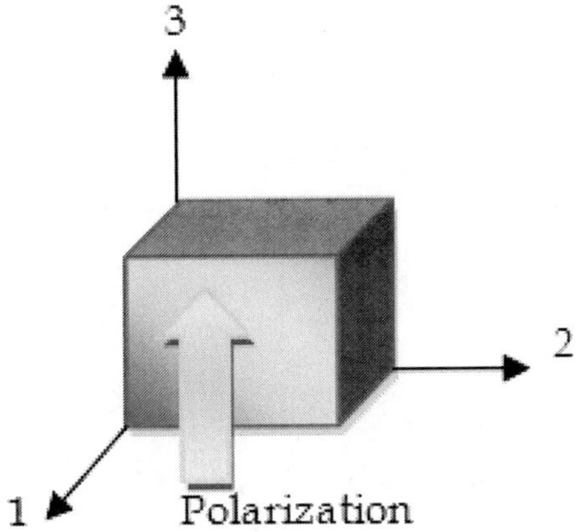

Figure 3. Defining the modes of piezoelectric material

The generated electrical charge of a piezoelectric material under a mechanical stress can be formulated in terms of dielectric displacement, *D* (charge per unit area, C/m^2) [Schwartz, 2002;Granstrom et al., 2007; Swallow et al., 2008]

$$D_i = d_{ij}\sigma_j \tag{1}$$

where *"d ij"* is the piezoelectric charge coefficient (C/N) and *"σ j"* is the stress (N/m^2). *"I"* is the direction of polarization and takes terms 1–3 and *"j"* is the direction of applied stress having subscripts 1–6 (Fig. 3).

In Figure 3, numbers 1, 2 and 3 present x,y and z axes of piezoelectric material, respectively. Modes 31 and 33 are two coupling modes of piezoelectric materials that when a force is applied in the perpendicular direction to the poling direction, the piezoelectric voltage coefficient acts in the mode 31. When a force is applied in the parallel direction to the poling direction, the piezoelectric voltage coefficient acts in the mode 33 which generally yields a higher coupling coefficient (*k*) [Roundy et al., 2005; Baker et al., 2007; Anton & Sodano 2007; Patel et al., 2010] which is the ability of piezoelectric materials to convert mechanical energy into electrical energy and vice versa, and can be defined by,

$$k = \sqrt{\frac{Y}{\varepsilon}}d \qquad (2)$$

where "ε" is dielectric constant and "*Y*" the elastic modulus of the material.

Piezoelectric voltage coefficient (*g*) relating the electric field generated by an applied mechanical stress and relationship between the piezoelectric charge coefficient and piezoelectric voltage coefficient can be expressed as [Jordan & Ounaies, 2001],

$$gij = dij / \varepsilon_0 \mathrm{Kj} \qquad (3)$$

where; "ε0" is the permittivity of free space (8.85x10-12 F/m) and "Kj" is the relative dielectric constant of the material.

Since the discovery of piezoelectricity in ceramics [Shirane & Suzuki, 1952; Jaffe et al., 1971] and polymers [Kawai, 1969], various studies have been carried out on structural changes [Ramos et al., 2005; Sencadas 2006], poling [Seo et al., 1985; Holstein et al., 1999; Neagu et al., 1999; Parvanova & Nadoliisky 2005] and applications, such as sensors [Tzou & Tseng 1990; Sirohi & Chopra 2000], actuators [Baz & Poh 1988; Schmidt et al., 2006], energy harvesting [Sodano & Inman 2004; Shu & Lien 2006; Granstrom et al., 2007; Ramadass & Chandrakasan 2010] and so on. PZT has been pre-eminent due to its piezoelectricity among other piezoelectric materials, with a piezoelectric coefficient (d_{33}) of 220pC/N [Hellwege, 1996] while PVDF exhibits much lower piezoelectric coefficient of $d_{33} \approx$ 35pC/N [Sencadas et al., 2006 ; De-Qing, 2008; Jain et al., 2010; Patel et al., 2010]. However, flexible nature of polymers adds extra versatility for applications against ceramic based piezoelectric materials.

With the invention of new polymers exhibiting piezoelectric and better mechanical properties, the scope of application has widened. Both ceramic and polymer based piezoelectric materials have found a wide range of application in many areas. Ceramic based piezoelectric materials, in general, have a higher piezoelectric constant compared to polymer based piezoelectric materials. However, polymer based piezoelectric materials have an advantage of being flexible which makes them preferable for many applications, particularly for wearable applications. This flexibility can also result in better conversion of energy in certain applications.

3.1. Flexible Piezoelectric Fibres

There are many naturally occurring piezoelectric structures, such as cane sugar, tendon, silk, bones etc., but polymers are not natural piezoelectric materials. However, polymers such as polyvinylidene fluoride (PVDF), polyproplylene (PP), polyethylene terephthalate (PET), odd numbered polyamides (PA11, PA9, PA7, PA5) etc. can be made piezoelectric (Kawai, 1969; Newman et al. 1980; Dunn & Carr, 1988; Harrison & Ounaies, 2001).

Siores et al (2010) were the first to develop and patent a continuous process for making piezoelectric fibres by extruding suitable polymers. To produce piezoelectric PVDF fibres via a continuous process on the melt extruder, granular PVDF was fed to extruder screw which was heated above melting point of the polymer. To gain the piezoelectricity, molecular chain of the PVDF fibre must be re-oriented and transformed from non-piezoelectric α-phase to piezoelectric β-phase by applying a high stretching ratio (Sencadas et al., 2006), heat (Neagu et al., 1999) at the stretching region and high voltage (Holstein et al., 1999& Ramos et al., 2005), simultaneously.

3.1.1. Fibre Extrusion and Poling

High purity PVDF polymer granules are fed a melt extruder. The extrusion temperature is kept at 195° C which is 20° C higher than the melting point of PVDF inside the feeding screw. The temperature is slightly higher at the die, 205° C, where the fibre is extruded. The extruded fibre is then air cooled with a blower and water cooled on the initial stage rollers which help in further cooling of the extruded fibre.

Poling is a critical step for piezoelectric fibre generation. Temperature, drawing ratio and applied electric field play a crucial role in the amount of polarisation. Highest polarisation charge coefficient was given in the literature (Sessler, 1981; Wegener et al., 2002).

The drawing of fibres takes place at the rollers, which have heating coil inside to vary the temperature during stretching of fibres. The temperature of these rollers is maintained constant on PVDF fibre when it leaves the roller and an appropriate electric field applied on PVDF fibre while being drawn.

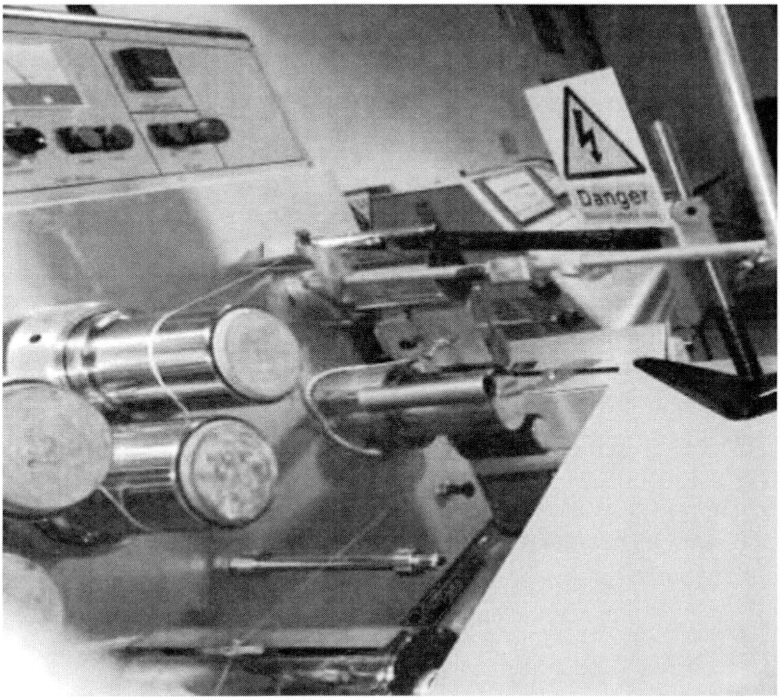

Figure 4. Piezoelectric polyvinylidene fluoride (PVDF) filament production via a continuous process using a melt extruder

Figure 4 shows the continuous process of producing piezoelectric polymer in a customised melt extruder. This is a less expensive and less time consuming method for preparing piezoelectric polymer fibres in that all process variables are applied simultaneously. Detailed information on polymer based piezoelectric fibre production via a continuous process has been reported (Siores et al, 2010).

3.1.2. Testing Of Generated Piezoelectric Fibres

Generated polymer fibres, shown in Figure 5, are embedded in between two thin sheets of aluminium or copper which act as electrodes. The fibres are placed close to each other such that the top electrode would not contact the bottom one. The top and bottom electrodes act as positive and negative terminals for the energy generating polymer piezoelectric device.

Poled PVDF fibres generate about 5 V_{oc} when a moderate mechanical stimulus is applied on to the fibres. The obvious advantage of producing flexible piezoelectric fibres is to be able to produce large area active surfaces by incorporating piezoelectric fibres in wearable technologies. However, to generate enough electricity for wearable applications to power small electronic devices, produced flexible piezoelectric fibres need to be used in a fabric structure such as woven, knitted, nonwoven and 3D structures.

Figure 5. Polymer based flexible piezoelectric fibre generated by a continuous process using the melt extruder.

4. SMART WOVEN FABRICS

Each fabric making method has its own special attributes which help us to find most applicable fabric structure and method for a specific application. By weaving polymer based piezoelectric fibres into woven fabrics, smart piezoelectric fabrics can be produced and used for many responsive applications.

Weaving is one of the best fabric making techniques that can be used for smart fabric production. Warp threads can be located at the heddles with different orders and wefts are travelled through warps by shuttle(s). The position of heddles designates where wefts will be going over or under the warps.

While a fabric is being designed, expectations from the final fabric are taken into consideration. For smart piezoelectric fabrics, depending on expected energy generation from the final product, weaving designs can be variable. In this case, intersection of piezoelectric (the charge generator) and conductive (the charge carrier) fibres is crucial. One piezoelectric fibre can interlace more than one conductive fibre. One conductive fibre can also interlace more than one piezoelectric fibre. However, one conductive fibre can only interlace the same pole of the each piezoelectric fibre.

A number of weaving designs are studied below for smart woven fabrics. Conductive fibres and conventional (non-conductive) fibres are needed alongside piezoelectric fibres. Because piezoelectric fibres carry negative charges on one side along its length and positive charge on the other side, a conductive material is needed to carry the charge produced by movements of the piezoelectric fibres. Conductive wires would add extra rigidity to the fabric which is an undesirable outcome for most textile structures.

The best alternative to undesirable wires may be conductive fibres are produced and patented (Perera & Mauretti, 2009). It is claimed (Mauretti & Perera, 2010) that conductive filaments are flexible, non-toxic and conformable for wearable applications. Electrical conductivity of metallised synthetic (acrylic) conductive textile yarns is widely studied (Vassiliadis et al., 2004, 2009, 2010). Mechanical and electrical properties of metallised conductive yarn are controlled by blending conventional and conductive fibres in the yarn and changing the ratio of fibres in the blend. The way piezoelectric, conductive and conventional fibres are integrated into fabric structure by weaving technique, gives a good indication of the performance of resultant fabric. When more piezoelectric fibres are used in the fabric, this results in higher energy generation by movement and mechanical strain. However, to be able to carry as much charge as it is possible, the right number of conductive fibres need to interlace with piezoelectric fibres.

The possible woven fabric designs for energy generation for wearable textiles are shown in this chapter. Blue lines represent piezoelectric fibres while red lines represent conductive and grey lines show non-conductive conventional fibres. This is the simplest weaving pattern produced by plain weaving technique. However, by integrating piezoelectric and conductive fibres into this basic structure, the resultant woven fabric becomes a smart fabric which can harvest energy from the natural sources.

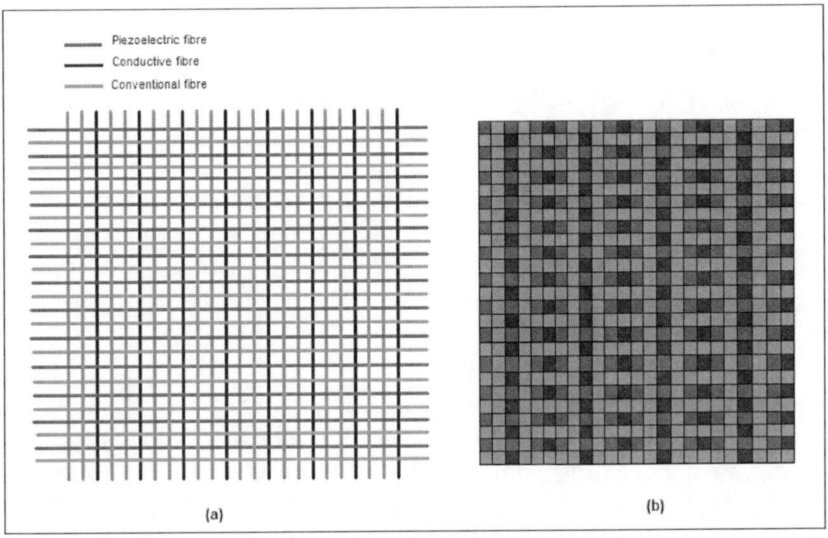

Figure 6. a) Smart woven fabric design 1 consisting of piezoelectric, conductive and non-conductive conventional fibres, (b) Face of the woven fabric consisting of piezoelectric, conductive and non-conductive fibres

Polymer based piezoelectric fibres can be used as either weft or warp into the woven structure and conductive fibres can be used as negative and positive electrodes for charge transfer so that the resultant fabric can produce energy for micro powered electronics.

The main advantage of the use of polymer based piezoelectric material in this application is its flexibility and the fact that it can easily be incorporated in the woven structures without causing any problem. It is impossible to integrate existing ceramic based piezoelectric fibres into similar structures because these fibres are rigid and brittle thus can cause major problems in the weaving process. For the first design shown in Figure 6(a), 2 heddles are needed to locate conductive and non-conductive fibres/yarns and 2 shuttles, the one with piezoelectric fibres/yarn and the other with non-conductive conventional fibres/yarn. In the warp direction, 2 conventional fibres are located between conductive fibres. Conductive fibres act as negative and positive electrodes.

If a number is given to each warp from left to right, odd numbered warps are located on the first heddle and even numbered warps are located on the second heddle. During the shuttles' travel along the loom's width, according to design, while the first conductive fibre only interlaces with negative pole of piezoelectric wefts, second conductive warp interlaces only positive pole of the piezoelectric filling fibres/yarns. Thus, any short circuit is avoided.

Figure 6(b) shows interlace of warp and weft threads and possible appearance on face of the fabric. If the used fibres counts are the same and the warps and wefts are located with an exact sequence, the resultant fabric will contain 24% piezoelectric, 16% conductive and 60% non-conductive conventional fibres/yarns.

The design shown in Figure 7(a) needs 2 heddles to locate conductive and non-conductive fibres/yarns and 2 shuttles, the one with piezoelectric fibres/yarn and the other with non-conductive conventional fibres/yarn. If a number is given to each warp from left to right, odd numbered warps are located on the first heddle and even numbered warps are located on the second heddle.

During the shuttles travel along the loom's width according to the design, the first heddle is kept in place, second heddle is uplifted so that warps are kept apart and shuttle travels through easily. Shuttle carrying piezoelectric fibre travels twice and then the other shuttle which carries non-conductive conventional fibres/yarn travels once. The whole process is repeated until the desired fabric structure is created. Thus, the first conductive warp only interlaces with negative charged sides of piezoelectric wefts, second conductive warp interlaces only with the positive charged sides of the piezoelectric filling fibres/yarns.

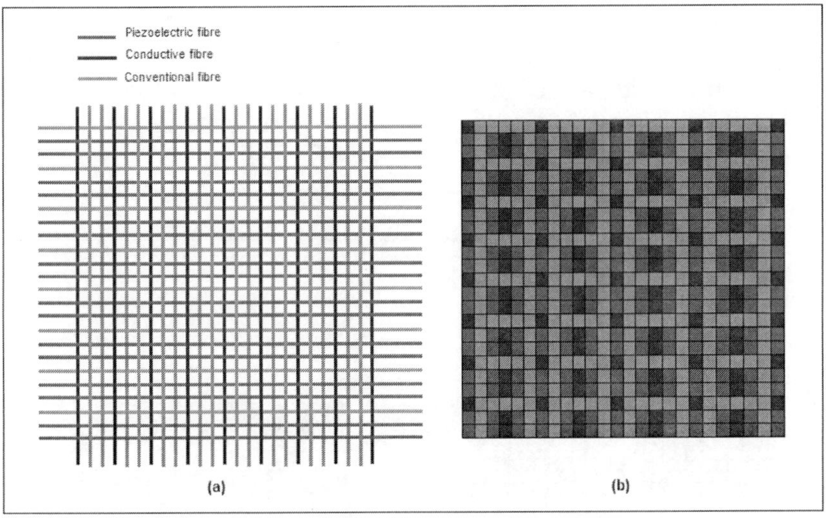

Figure 7. a) Smart woven fabric design 2 consisting of piezoelectric, conductive and non-conductive conventional fibres, (b) Face of the woven fabric consisting of piezoelectric, conductive and non-conductive fibres

Figure 7(b) shows interlace of warp and weft threads and possible appearance on face of the fabric. If the used fibres counts are the same and the warps and wefts are located with an exact sequence, the resulted fabric will contain 34% piezoelectric, 18% conductive and 48% non-conductive conventional fibres/yarns.

The design shown in figure 8(a) needs 2 heddles to locate conductive and non-conductive fibres/yarns and 2 shuttles, the one with piezoelectric fibres/yarn and the other with non-conductive conventional fibres/yarn. If we give a number to each warp from left to right, 1st, 2nd, 7th, 8th, 13th, 14th, 19th, 20thand 25th warps are located on the first heddle and other warps are located on the second heddle.

According to design in figure 8(a), while first heddle is kept in place, second heddle is uplifted so that warps can be kept apart from the first heddle's warps and shuttle, which carries piezoelectric fibres/yarn, can easily travel through. The shuttle which carries piezoelectric fibre travels twice and then the first heddle is uplifted while the second heddle is lowered so that the other shuttle which carries non-conductive conventional fibres/yarn travels once through the warps. The same movements are carried out with the same order again and again until a fabric structure is created. Thus, all the conductive warps on the first heddle only interlace with negative pole of piezoelectric wefts and all the conductive warps on the second heddle interlace only with positive pole of the piezoelectric wefts.

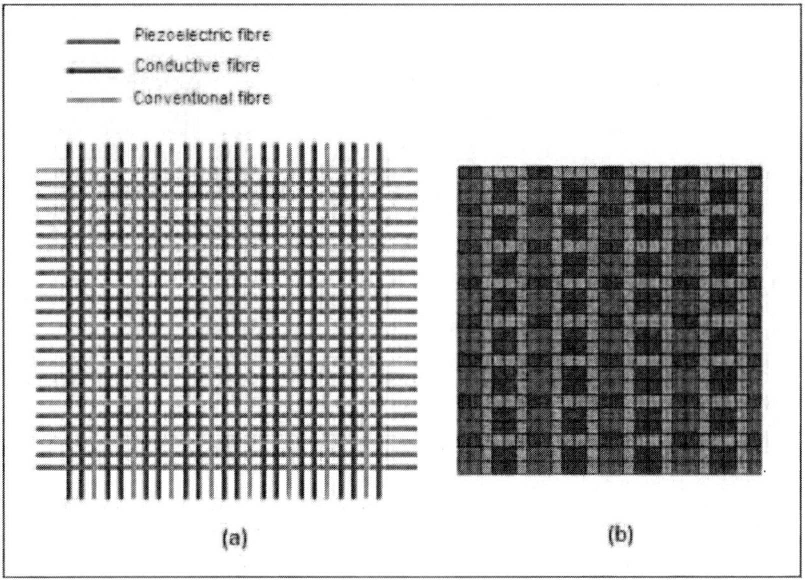

Figure 8. a) Smart woven fabric design 3 consisting of piezoelectric, conductive and non-conductive conventional fibres, (b) Face of the woven fabric consisting of piezoelectric, conductive and non-conductive fibres

Figure 8(b) shows interlace of warp and weft threads and possible appearance on face of the fabric. If the used fibres' counts are the same and the warps and wefts are located with an exact sequence, the resultant fabric will contain 34% piezoelectric, 34% conductive and 32% non-conductive conventional fibres/yarns.

The design shown in Figure 9(a) needs 2 heddles to locate conductive and non-conductive fibres/yarns and 2 shuttles, the one with piezoelectric fibres/yarn and the other with non-conductive conventional fibres/yarn. If we give a number to each warp, 1st, 2nd, 5th, 6th, 9th, 10th, 13th, 14th, 17th, 18th, 21st, 22nd and 25th warps are located on the first heddle and others are located on the second heddle.

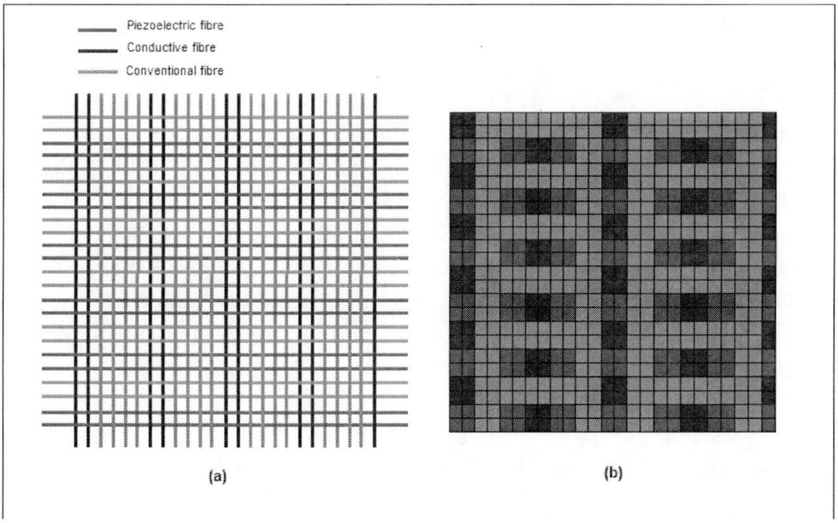

Figure 9. a) Smart woven fabric design 4 consisting of piezoelectric, conductive and non-conductive conventional fibres, (b) Face of the woven fabric consisting of piezoelectric, conductive and non-conductive fibres

During the first shuttle's travel along the loom width, the first heddle is kept in place and the second heddle is uplifted so that warps can be kept apart from the first heddle's warps and shuttle which carries piezoelectric fibres/yarn can easily travel through. The shuttle carrying piezoelectric fibre travels twice and then the first heddle is uplifted while the second heddle is lowered so that the other shuttle which carries non-conductive conventional fibres/yarn travels twice through the warps. The same movements are carried out in the same order again and again until a fabric structure is created. Thus, all the conductive warps on the first heddle only interlace with negative pole of piezoelectric wefts and all the conductive warps on the second heddle interlace only with positive pole of the piezoelectric wefts.

Figure 9(b) shows interlace of warp and weft threads and possible appearance on face of the fabric. If the used fibres counts are the same and the warps and wefts are located with an exact sequence, the resultant fabric will contain 26% piezoelectric, 18% conductive and 56% non-conductive conventional fibres/yarns.

5. CONCLUSIONS

Polymer based piezoelectric fibres can be used as either weft or warp into the woven structure and conductive fibres can be used as negative and positive electrodes for charge transfer, therefore the resultant fabric can produce energy to power small electronic devices.

The advantage of polymer based piezoelectric fibre is their flexibility so that they can easily be used in woven structures. It is impossible to integrate existing ceramic based piezoelectric fibres into a similar structure since they are brittle.

For all four fabric designs studied in this chapter, interlace of piezoelectric and conductive fibre/yarn is significant. In a woven fabric structure, one piezoelectric fibre can interlace with more than one conductive fibre and one conductive fibre can also interlace with more than one piezoelectric fibre. However, to avoid any short circuit, one conductive fibre can only interlace with the same pole of the each piezoelectric fibre.

Since the fibres are considered having the same thickness, in the first design, piezoelectric and conductive fibres interlace 96 times in the fabric. The times of interlace of piezoelectric and conductive fibres are 153 for the second design whilst it is 289 and 117 times for the third and forth fabric designs, respectively. Therefore, the highest energy generation is expected from the third design when they all designs are subjected to the same amount of mechanical stimulus.

Smart piezoelectric woven fabrics can be used where they can be subjected to mechanical strain/stress or vibrations. Depending on the application and energy need, smart piezoelectric woven fabrics can be used to produce whole textile structure or only a part of it. For instance, tents, awnings and umbrellas can be wholly made of smart piezoelectric fabrics and produce electricity under rain as well as wind. However, waterproof finishing is needed if the fabric will be used for outdoor applications.

Energy generated by piezoelectric materials is always in the form of AC, therefore a small rectifier is needed for the conversion of the generated energy (AC) into usable energy (DC) for low power electronics.

The best weaving technique for smart piezoelectric fabric is plain weaving and its derivatives. Possible smart woven fabric designs are not limited. Depending on energy need, different fabric designs can be made with less or more interlacing. Other existing fabric making methods (other than weaving) such as embroidery can also be used to produce smart fabrics.

REFERENCES

1. S. R. Anton, H. A. Sodano, 2007 Smart Mater. Struct., 16 R1-R7.

2. J. Baker, S. Roundy, P. Wright, 2005 Proc. 3rd Int. Energy Conversion Engineering Conf., 959 970.

3. A. Baz, S. Poh, 1988 Performance of an active control system with piezoelectric actuators, J. Sound and Vibration, 126 2 327 343.

4. Eric. Broudy, 1979 The Book of Looms: A History of the Handloom from Ancient Times to the Present, University Press of New England, 0-87451-649-8 York, USA

5. Z. De -Qing, 2008 Chinese Physics Letters, 25 4410

6. P. E. Dunn, S. H. Carr, 1988 Piezoelectricity in nylon 5,7, IEEE Proceedings 6th International Symposium on Electrets, DOI 10.1109/ISE.1988.38559

7. D. A. Gioello, 1982 Understanding Fabrics from Fiber to Finished Cloth, Fairchild Publications, 870053779 York, USA

8. J. Granstrom, J. Feenstra, H. A. Sodano, K. Farinholt, 2007 Energy harvesting from a backpack instrumented with piezoelectric shoulder straps, Smart Materials Structures, 16 1810 1820.

9. J. S. Harrison, Z. Ounaies, 2001 Piezoelectric Polymers, NASA/CR-2001 211422 , ICASE Report 2001-43

10. K. H. Hellwege, A. M.. Hellwege, Eds 1996 Ferroelectrics and Related Substances: Oxides, Landolt Bornstein New Series, Springer, Berlin.

11. P. Holstein, N. Leister, U. Weber, D. Geschke, H. Binder, 1999 A Combined Study of Polarization Effects in PVDF, 10th International Symposium on Electret. IEEE

12. B. Jaffe, W. R. Cook, H. Jaffe, 1971 Piezoelectric Ceramics, Academic Press, London.

13. A. Jain, J. S. Kumar, D. R. Mahapatra, H. H. Kumar, 2010 Proceedings of SPIE- The International Society for Optical Engineering 7647.

14. T. L. Jordan, Z. Ounaies, 2001 NASA ICASE Report 2001-28 2001 28.

15. H. Kawai, 1969 The piezoelectricity of poly(vinylidene fluoride), Jpn. Journal of Applied Physics, 8 975 976

16. L. Mateu, F. Moll, 2005 J. Intell. Mater Syst. Struct., 16 835 845.

17. G. J. Mauretti, W. R. Perera, 2010 iCon Conductive Filaments, Proceeding of Textile Institute Centenary Conference, Manchester, UK, November 3-4, 2010

18. E. R. Neagu, J. S. Hornsby, D. K. Das-Gupta, 1999 Analysis of Polarization ans Space Charge in Thermally Poled PVDF, 10th International Symposium on Electrets, IEEE, 87 90

19. B. A. Newman, P. Chen, K. D. Pae, J. I. Scheinbeim, 1980 Piezoelectricty in Nylon 11, Journal of Apllied Physics, 51 10 5161 5164 , DOI 10.1063/1.327283

20. V. D. Parvanova, M. M. Nadoliisky, 2005 Polarization Processes in PZT Ceramics, Bulg. J. Phys., 32 45 50.

21. W. R. Perera, G. J. Mauretti, 2009 Microwires, Methods for Their Production, and Products Made Using Them, US Patent 20090260848 A1

22. Y. K. Ramadass, A. P. Chandrakasan, 2010 An Efficient Piezoelectric Energy Harvesting Interface Circuit Using a Bias-Flip Rectifier and Shared Inductor, IEEE Journal of Solid-State Circuits, 45 1 189 204.

23. M. M. D. Ramos, H. M. G. Correia, S. Lanceros-Mendez, 2005 Atomistic modelling of processes involved in poling of PVDF, Computational Materials Science, 33 230

24. S. Roundy, P. K. Wright, J. Rabaey, 2003 Computer Communications, 26 1131 1144.

25. V. H. Schmidt, L. Lediaev, J. Polasik, 2006 Piezoelectric Actuators Employing PVDF Coated with Flexible PEDOT-PSS Polymer Electrodes, IEEE Transactions on Dielectrics and Electrical Insilation, 13 5 1140 1148.

26. M. Schwartz, 2002 Encyclopaedia of Smart Materials 1-2, 780 792.

27. V. Sencadas, V. M. Moreira, S. Lanceros-Mendez, A. S. Pouzada, R. Gregorio Jr, 2006 α-to-β Transformation on PVDF Films Obtained by Uniaxial Stretch, Materials ScienceForum 514-516, 872 876

28. V. Sencadas, R. G. Filho, S. Lanceros-Mendez, 2006 J. Non Cryst. Solids 352 2226

29. J. W. Seo, K. S. Ryoo, H. S. Lee, 1985 Characteristics of Charge Traps and Poling Behaviour of Poly(Vinylidene Fluoride), Bulletin of Korean Chemical Society, 6 4 218 221.

30. G. M. Sessler, 1981 Piezoelectricity In Polyvinylidene fluoride, J. Acoust. Soc. Am., 70 1596 1608.

31. G. Shirane, K. Suzuki, 1952 Crystal structure of Pb(Zr-Ti)O3, J. Phys. Soc. Jpn., 7

32. Y. C. Shu, I. C. Lien, 2006 Analysis of Power Output for Piezoelectric Energy Harvesting Systems, Smart Materials Structures, 15 1499 1512.

33. E. Siores, R. L. Hadimani, D. Vatansever, 2010 Piezoelectric Polymer Element & Production Method & Apparatus Therefor. GB Patent Application 1015399 7

34. J. Sirohi, I. Chopra, 2000 Fundamental Understanding of Piezoelectric Strain Sensors, J. Intelligent Mater. Systems and Struc., 11 246 257.

35. H. A. Sodano, D. J. Inman, G. Park, 2004 Shock Vib. Digest, 36 197 205.

36. H. A. Sodano, D. J. Inman, 2004 A Review of Power Harvesting from Vibration Using Piezoelectric Materials, The Shock and Vibration Digest, 36 3 197 205.

37. L. M. Swallow, J. K. Luo, E. Siores, I. Patel, D. Dodds, 2008 Smart Mater. Struct., 17

38. H. S. Tzou, C. I. Tseng, 1990 Distributed piezoelectric sensor/actuator design for dynamic measurement/control of distributed parameter systems: A piezoelectric finite element approach, J. Sound and Vibration, 138 17 34.

39. M. Umeda, K. Nakamura, S. Ueha, 1997 Jpn. Journal of Applied Phys., 36 3146 3151.

40. S. Vassiliadis, Ch. Provatidis, K. Prekas, M. Rangoussi, 2004 Electrically Conductive Spun Yarns, Proc. of the Xth International Izmir Textile and Apparel Symposium, 37 49 , Izmir, Turkey

41. S. Vassiliadis, M. Rangoussi, D. Meimaris, K. Prekas, 2009 Provaditis Ch.Electrically Conductive Spun Yarns and their Contact Behaviour, Proceedings of the International Conference on Intelligent Textiles and Mass Customisation, Casablanca, Morocco

42. S. Vassiliadis, K. Prekas, M. Rangoussi, K. Absalon, J. Maillard, 2010 The Conductive Spun Yarns as Electrical Components, Proceedings of the XIIth International Izmir Textile and Apparel Symposium, 333 338 , Izmir, Turkey

43. M. Wegener, W. Künstler, K. Richter, R. Gerhard-Multhaupt, 2002 Ferroelectric polarization in stretched piezo- and pyroelectric poly(vinylidene fluoride-hexafluoropropylene) copolymer films, J. Appl. Phys., 92 7442 7447.

44. Wikipedia, online access from http://en.wikipedia. org/wiki/Loom# Handloom (Jan. 2011)

Chapter 2

Electro-Conductive Sensors and Heating Elements Based on Conductive Polymer Composites in Woven Fabric Structures

Irina Cristian[1, 2, 3], Saad Nauman[1, 2], Cédric Cochrane[1, 2], Vladan Koncar[1, 2]

[1] University Lille Nord de France, France

[2] ENSAIT, GEMTEX, F-59100 Roubaix, France

[3] Technical University GHEORGHE ASACHI of Iaşi, Romania

1. INTRODUCTION

The use of intelligent materials which are capable of reacting to external stimuli is growing in the field of textiles. Conductive materials, metals as well as conducting polymers, are already being used in many textile applications, like antistatic materials, electromagnetic interference shielding, heating, transport of electrical signals, sensors etc.

Conductive polymers are some of the developments which seem to respond to specific properties of textile materials, like flexibility and deformability. Two sub-classes of conductive polymers can be identified: intrinsically conductive polymers and conductive polymer composites.

Inherently conducting polymers are composed of polymer chains containing long conjugated double bonds which give rise to highly conducting properties, and they have been considered as promising materials (Heeger, 2002; Kumar et al., 1998). Inherently conducting polymers are suitable for applications in many domains of intelligent textiles, but they present some substantial disadvantages, like infusibility or insolubility in common organic solvents, weak mechanical properties and poor procesability. However, advanced solution processing of this class of polymers has been developed significantly over the last decade to improve their low solubility. For example, chemical modification of monomers with dopants has enhanced the solubility in the case of polythiophene and polyaniline (Haba et al., 1999; Gettinger et al., 1995).

For their part, composite conductive polymers are obtained by blending (generally by melt mixing) an insulating polymer matrix (thermoplastic or thermosetting plastic) with conductive fillers like carbon black, carbon fibres or nanotubes, metallic particles or conductive polymers. The presence of filler particles in the matrix may have a negative impact on the mechanical properties of the final composite (Krupa et al., 2001; Novak et al., 2002). Instead of this, the development in the field of composite conductive polymers seems therefore to be a promising approach for intelligent textiles own to simplicity of preparing and to their low cost.

In this chapter, two applications (sensor and actuator) based on coating of textile materials with conductive polymer composites are presented. Several different coating techniques for intelligent textile structures exist - the one chosen for both applications was developed in our laboratory (Cochrane et al., 2007, 2010) based on dispersed carbon black particles (Printex® L6) in a polymer solution (Styren-Butadien-Styren or latex), using a solvent.

In the first part of this chapter, a new approach of NDE (Non Destructive Evaluation) using fibrous sensors inserted inside composite woven reinforcements during their weaving is presented. The use of 3D woven fabrics as the reinforcing medium for composites is becoming a popular choice, due to various advantages such as reduced cost and shorter production cycle, greater design flexibility and superior mechanical properties (Kamiya et al., 2000). Recently, these high performance composites reinforced with 3D structures have found wide applications in various industrial areas such as aerospace, aircraft, automobile, civil engineering etc. Good quality and reliability are basic requirements for advanced composite structures which are often used under harsh environments. To improve their performance, the cure monitoring of technological process is clearly necessary. At the same time, in service NDE is also needed to keep these structures operating safely and reliably.

A novel flexible piezoresistive fibrous sensor has been developed and optimized for in situ structural deformation sensing in carbon composites. Those sensors were inserted as weft in 3D-woven interlock reinforcement, during the weaving process on a special weaving loom. The reinforcement was then impregnated in epoxy resin and was later subjected to quasi static tensile loading. It was found that the sensor was able to detect deformations in the composite structure until rupture since it was inserted together with reinforcing tows.

The morphological and electromechanical properties of the fibrous sensors have been analyzed using tomography and yarn tensile strength tester. An appropriate data acquisition module has also been developed and used for data acquisition and its further treatment. The results obtained for carbon composite specimens under standard testing conditions have validated in situ monitoring concept using our fibrous textile sensors.

The second part of this chapter presents a woven fabric containing an original heating element. Textile actuators like heating fabrics can find applications in numerous and varied fields such as sports, leisure, medical and automotive (Droval et al., 2005; El-Trantawy et al., 2002). Usually actuators need heavy power supplies that are rarely flexible and lightweight which badly affects wearability. Our heating element is designed to adapt to flexible structures. Metallic yarns, used as electrodes are integrated in a woven structure in a comb-teeth arrangement. These electrodes are connected to a power supply. A thin conductive coating is applied on the fabric surface and on the electrode arrangement, in order to ensure uniform heat distribution. The coating is a composite material based on aqueous latex dispersed with carbon black as filler. The heating element (comb electrodes and electro-conductive coating) can thus adopt the desired pattern. This is an important aspect of our heating element as it allows integration of the heating element in various fabrics designed for varied and diverse applications. In our research the distance between electrodes remained unchanged, the only variable value being the percentage of filler. The results regarding the power consumption and textile surface temperature for different filler content were registered. From these results, optimum filler content for heating application was defined. A thermal image of the heating fabric operating at low voltage is given, in order to demonstrate the efficiency of our system. A more precise measurement of surface temperature is thus possible. For applied voltage of 15 V, the maximum temperature gained was about 50 °C. The thermal image also demonstrates the homogeneity of heating provided by our system. Potential applications of these self heating fabrics include garments designed to provide thermal comfort and antifreeze safety.

2. ELECTRO-CONDUCTIVE SENSORS FOR ON-LINE MEASUREMENTS OF STRUCTURAL DEFORMATION IN COMPOSITES REINFORCED WITH 3D-WOVEN FABRICS

Weaving technique has been used for a long time in order to obtain technical textile products for industrial applications. An important use of this technology is for the manufacturing of 3D reinforcements using high performance fibres (carbon, glass, aramid etc.). 3D reinforcement based composites, in combination with high-performance fibres, are being increasingly used in the aerospace industry (Ko, 2007). Particular advantages of these fabrics mentioned in the literature include better through the thickness properties, better out of plane properties, high impact resistance, enhanced delamination resistance, resistance to crack propagation, impact/fracture resistance, improved post impact mechanical properties, damage tolerance, dimensional stability, ease of fabrication and minimal need of cutting, lay up and joining.

Regarding these properties it can be safely concluded that 3D woven fabrics constitute the most promising class of reinforcements for composite materials for high tech structural applications. To improve their performance, the cure monitoring of technological process is clearly necessary. At the same time, in-service non destructive evaluation is also needed to keep these structures operating safely and reliably. Non destructive evaluation techniques have been developed in the past including ultrasonic scanning, acoustic emission, shearography, stimulated infrared thermography, fibre brag grating sensors and vibration testing etc. (Black, 2009). The challenge today is to develop new low cost techniques which can perform on-line structural health assessment starting from the manufacture of composite structure to the real service of these structures in the field. Moreover, the non destructive evaluation techniques have to be integrated in the design phase and sensors should be inserted during the fabrication of composites, in order to improve accuracy and reduce their costs. The classical non destructive evaluation techniques are difficult to adapt. They are not well suited for on-line structural health monitoring, because of difficulties in making in situ implementation.

One possible solution is to use intelligent textile materials and structures which provide real possibility for on-line and in situ monitoring of structural integrity. Such intelligent materials are made by coating or treating textile yarns, filaments or fabrics with nanoparticles or conductive & semi-conductive polymers giving them special properties.

A review of piezoresistive sensing approaches already being applied to measure strain in fabrics/composites shows that several sensing mechanisms exist (Dharap et al., 2004; Lorussi et al., 2005; Scilingo et al., 2003; Fiedler et al., 2004). These approaches may be categorized on the basis of manufacturing technology as nanotube networks, use of carbon tows for self-sensing and semi-conductive coatings.

Nanotubes have been investigated in detail for use as sensing mechanisms, both for smart textile applications and for structural health monitoring of composites. Significant challenges still exist in their development, for example the efficient growth of macroscopic-length carbon nanotubes, controlled growth of nanotubes on desired substrates, durability of nanotube based sensors and actuators, effective dispersion in polymer matrices and their orientation. Therefore, there is a need to develop both experimental and analytical techniques to bridge the nano and macro scales towards optimization so as to use nanotube networks as sensors inside macroscale (fabric) or mesoscale (tow) composites (Li et al., 2005).

Carbon fibre reinforced composites offer a unique possibility of using carbon tows as sensing network because of their conductivity. However such an approach can only be used for conductive fibre based composites. Moreover, before applying such an approach for structural health monitoring it is imperative to understand the deformation mechanism of the reinforcement. Any anomaly in the deformation mechanism can threaten the sensing mechanism's validity and efficacy.

Concerning semi conductive coatings, they are easy to realize and can be made wash resistant. Their use as percolation networks for sensing in structural health monitoring applications is quite promising and needs to be further investigated.

Present study is aimed at designing, developing and optimizing piezoresistive fibrous sensors realized from semi-conductive coatings, suited for composites structural parts containing 3D reinforcement. Our sensors can be embedded inside the reinforcement during weaving and they have all the characteristics of a traditional textile material (flexible, lightweight and are capable of adopting the geometry of the reinforcement and become its integral part).

Embedding such an intelligent piezoresistive sensor inside the reinforcement during weaving process is the most convenient and cost effective way of insertion of a sensor for structural health monitoring.

Development and optimization of such piezoresistive sensors has been carried out in order to render them sensitive enough to measure in situ strains inside the composite part. Sensitivity is important as the targeted application usually undergoes very low strains and even such low strains and/or vibrations during the life time of composite parts are critical. Often they are used in areas where structural integrity can not be compromised (aircraft wings, bodies etc).

2.1. Sensor Design and Optimisation

As coating solution, the conductive polymer composite based on dispersion of carbon black particles (Printex® L6) in polymer (Evoprene® 007) solution, using chloroform as a solvent was chosen (Cochrane et al, 2007, 2010).

In order to characterize the sensitivity and adherence of the coating on different substrates, the 35 % carbon black solution was coated on different yarns (71 tex cotton spun yarns; 482.3 tex polyethylene monofilament and 25 tex polyamide monofilament). Visual inspection of the surfaces of the coated yarns shows that the coating is more uniform for synthetic monofilaments compared to cotton yarns. The cotton yarns absorb the conductive solution, which penetrates inside the pores and interstices much like a dye. This particular phenomenon could be a source of non homogeneity in sensor electrical and mechanical properties, as the spun yarn is non uniform as compared to filaments, the coating and thus the resistivity achieved could be non uniform. Moreover the greatest inconvenience with coated cotton spun yarns is their low sensitivity during the initial tensile loading phase.

The resistance values were measured on 12 coated samples of each variant using a multimeter. The resistivities were then calculated using the yarn/filament fineness, yarn/filament lengths and these measured resistance values. Fig. 1 gives a comparison of calculated resistivity values of conductive layer deposited on different fibrous substrates. It can be seen that coated polyethylene filaments show relatively lower dispersion of resistivity as compared to coated poyamide filaments.

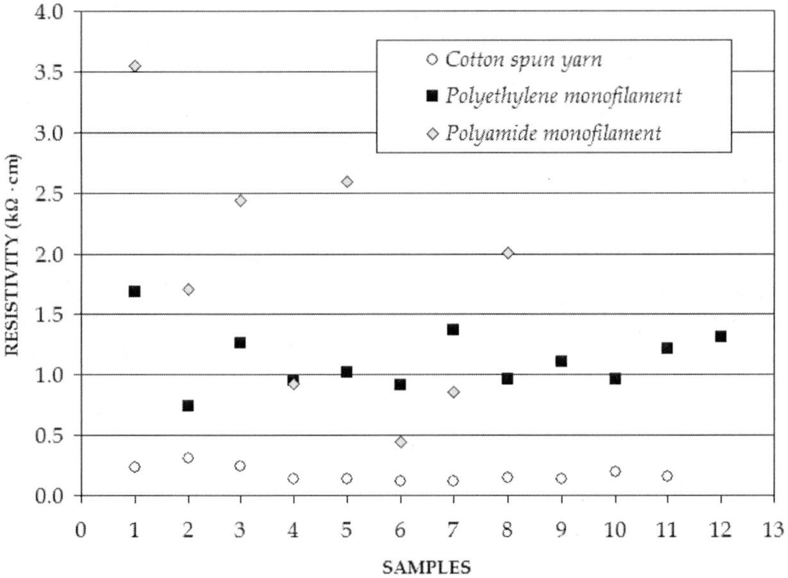

Figure 1. Resistivity values calculated for different substrates coated with 35 wt.- % carbon black solution

In order to carry out tensile tests on coated yarns and monofilaments, MTS 1/2 tester was used. Samples underwent quasi static tensile loading at a constant test speed of 5 mm/min. For the purpose of electrical resistance variation measurement during the tensile testing, a simple voltage divider circuit and Keithley KUSB-3100 data acquisition module were employed. Fig. 2 shows some of the results for electrical resistance variation, expressed as normalised resistance ($\Delta R/R$) during tensile testing, obtained using different substrates for coating.

Initial resistance of the coating on cotton yarns is much lower then monofilaments. But since the cotton spun yarns are inherently irregular, the coatings obtained are not homogenous and the results for different coated yarns vary widely in their response to tensile loading (Fig. 2-a). Due to particular fineness of the polyamide monofilament it was found that slight non homogeneity in coating on the surface can result in breakdown of conductive path as is obvious from Fig. 2-b. As a result, the behaviour of polyamide is highly inconsistent. Polyethylene monofilaments provide a reasonably good compromise as the substrate. The coatings on polyethylene are easy to achieve due to good substrate/conductive solution interfacial properties. As the curves in Fig. 2-c show, the polyethylene coatings are reproducible as the curves for all the four samples are nearly identical as opposed to polyamide and cotton. Therefore, polyethylene monofilament was chosen for sensor development.

The two ends of the coated polyethylene filaments were additionally coated with silver paint and fine copper wire was attached to the two ends with the help of this paint (as shown in the Fig. 3). In this way, secure connections were realized enabling the reduction of the contact resistance to the minimum.

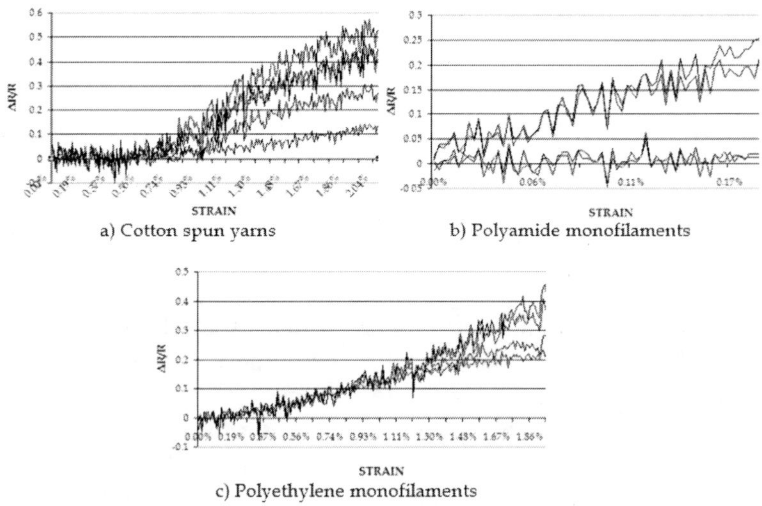

Figure 2. Electrical resistance variation during tensile strength tests on different yarn and filament substrates coated with 35 wt.-% carbon black solution

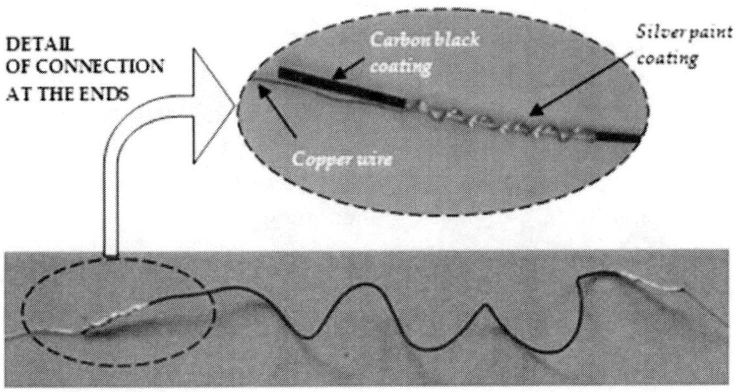

Figure 3. Carbon black coated sensor with polyethylene substrate

Sensor structural and geometrical parameters along with initial electrical resistance are shown in Table 1.

Table 1. Sensor properties

No.	PARAMATER	UM	VALUE
1	Linear density of the filament	g/km	48.23
2	Diameter of the filament	mm	0.70
3	Average width of the sensor cross section	mm	1.68
4	Average thickness of the sensor cross section	mm	1.26
5	Aspect ratio of the sensor (width/thickness)	-	1.33
6	Initial resistance of the sensor	kΩ	43.3

For insertion in conductive fibre based reinforcements like that woven using carbon multifilament tows, the sensor was coated with Latex Abformmasse supplied by VossChemie® so as to insulate the sensor from surrounding carbon tows.

Prepared in this way, the sensor with polyethylene substrate was tested again on MTS 1/2 tester, under quasi static tensile loading at a constant test speed of 5 mm/min. The same Keithley® KUSB-3100 data acquisition module was employed for the purpose of voltage variation during tensile testing. This time, a special set-up containing a Wheatstone bridge and an amplifier was used to measure unknown variable resistance of the sensor as a function of output voltage. As is obvious from curves presented inFig. 2-a, b and c, the simple voltage divider circuit is not adequate for the measurement of resistance change in case of sensors developed here. These piezoresistive sensors produce a very small percentage change in resistance in response to physical phenomena such as strain. Moreover the output signal has considerable noise. Generally, a bridge measures resistance indirectly by comparison with a similar resistance. Wheatstone bridges offer an attractive solution for sensor applications as they are capable of measuring small resistance changes accurately (Wilson, 2004).

Fig. 4 shows schematic diagram of the data acquisition module developed and used for data acquisition and its further treatment.

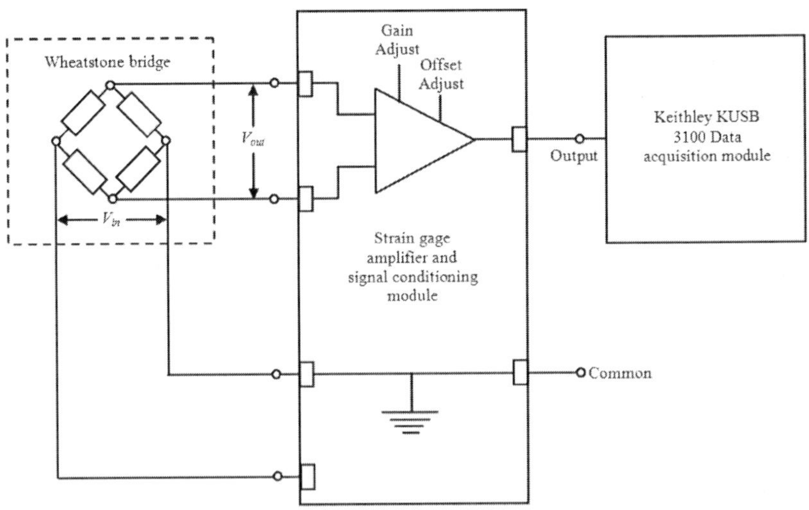

Figure 4. Schematic of instrumentation amplifier (INA 101) connected to Wheatstone bridge

The resistance variation data thus obtained for different test results was treated for noise reduction using a low pass filter. The resultant stress-strain-resistance relationship curve up to 2.75 % elongation of the out of composite sensor (before insertion in the reinforcement) is shown in Fig. 5.

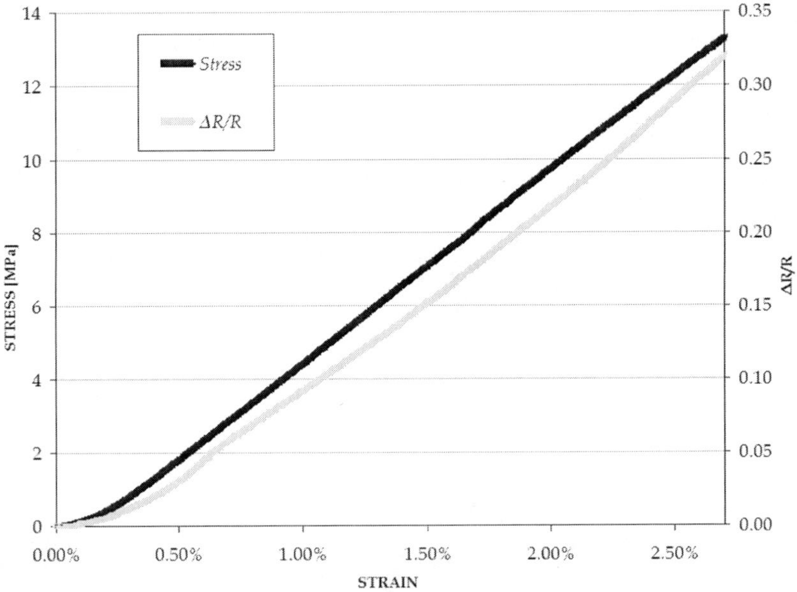

Figure 5. Normalized resistance and stress against strain for sensor outside composite

It may be noticed in Fig. 5 that the stress vs. strain curve has the same shape as normalised resistance (ΔR/R) vs. strain curve. This validates electromechanical properties of our fibrous sensor for strains ranging from 0 to 2.75 %.

In Fig. 6, the hysteresis results of the sensor for 10 cycles have been given. The sensor underwent 0.5 % extension at a constant test speed of 5 mm/min, followed by compression in each cycle. The sensor follows the extension and compression patterns in each cycle.

The hysteresis is high for the first cycle which reduces gradually and for the 10th cycle the sensor exhibits almost linear behaviour.

2.2. Sensor Insertion In Carbon Woven Reinforcement

An orthogonal/layer to layer warp interlock with 13 weft layers and 12 warp layers was chosen as woven structure (Fig. 7-a) and than was woven on a modified conventional loom (Patronic B60 ARM). 6K multifilament carbon tows (supplied by Hercules Inc.) having 200 tex was used in both direction – warp and weft. Yarn densities were 24 yarns/cm in warp direction and 170 yarns/cm in weft direction. The thickness and areal density of resulted reinforcement were 6.5 mm and 3908 g/m^2, respectively.

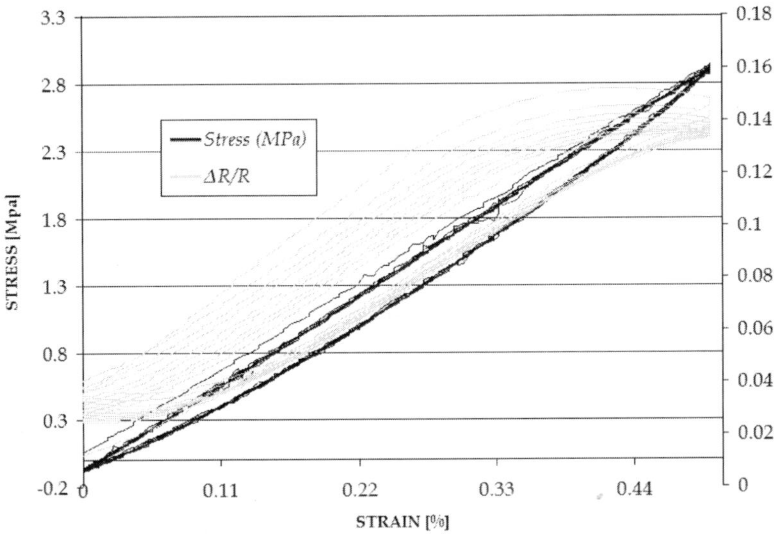

Figure 6. Normalized resistance (ΔR/R) and stress against strain for sensor (Hysteresis 10 cycles at 0.5 % extension)

Sensors can be inserted in warp or weft directions during weaving. Given the technical complications associated with sensor insertion in warp direction during weaving on a loom, insertion in weft direction has been carried out for preliminary studies. The placement of sensor in the reinforcement was decided so that the sensor was inserted in the middle of the structure related to thickness (Fig. 7-b).

The sensor was inserted during the weaving process, as a weft yarn and it follows the same trajectory as the carbon weft yarns inside the reinforcement. In Fig. 8, off the loom dry reinforcement photograph have been shown. Latex coated sensor connections can be seen protruding from the reinforcement.

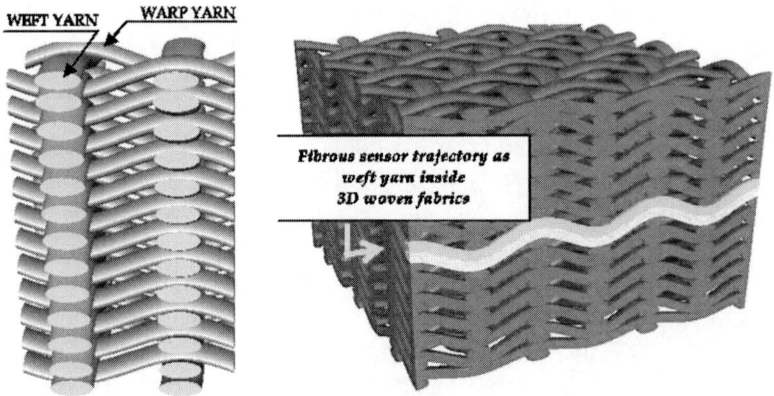

a) The weave repeat

b) The path of the sensor inside woven fabric

Figure 7. Interlock weave structure used as reinforcement – graphical representation (TexGen software)

Figure 8. Reinforcement with protruding sensor connections

2.3. Carbon Woven Reinforcement Impregnation and Testing

After weaving, the reinforcement was carefully removed from the loom and was impregnated using vacuum bag infusion process in order to make the composite part stiff. The resin employed was epoxy Epolam® 5015. The two connections of the sensor which remain outside the reinforcement at the two ends were carefully separated from the rest of the mould. This was done by creating two vacuum sub moulds inside the larger mould so that the resin may not impregnate the two connections of the sensor. The impregnated composite samples were cut into slabs of 25 X 2.5 cm (Fig. 9).

The composite specimens were tested on Instron 8500 tester. Tensile strength tests were performed on the composite specimens (according to ISO 527-4, 1997) in the weft direction i.e., the direction parallel to the inserted sensor. The same Wheatstone bridge was used for resistance variation measurement. The configuration of the testing equipment was also kept the same. The composite structural part was tested at constant test speed of 5 mm/min. The composite underwent traction until rupture.

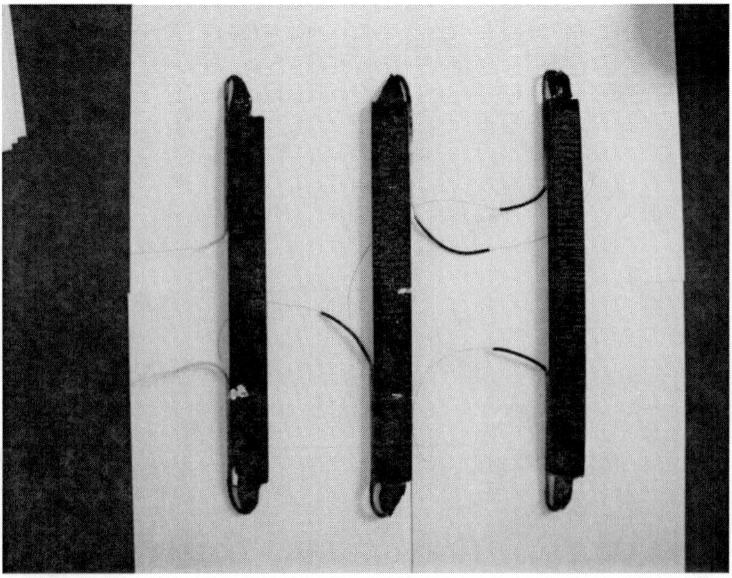

Figure 9. Textile composite sample containing fibrous piezo-resistive sensor

2.4. On-Line Measurements Of Sensor In Woven Fabric Composites - Results

Resultant stress-strain-resistance relationship curve is shown in Fig. 10. It can be observed that the normalized resistance follows the stress-strain curve. The stress-strain-resistance curve can be divided into four regions: *the initial stiff region* - where the composite exhibits toughness against the applied load represented by high slope; *the tows straightening region; the second stiff region* and the *zone of rupture*. The rupture occurred at the strain of 0.52 %, after which the tensile strength tester came back to its initial position at the same speed (5 mm/min). Since the fibrous sensor has not been broken, the normalised resistance (ΔR/R) decreased until zero as the tester returned to its initial position. However this decrease was not linear because the sensor was still intact while the resin-sensor interface was partially damaged which caused its non linear behaviour.

Due to the high difference in yarn densities (24 warp yarns/cm vs. 170 weft yarn/cm), the weft tows are highly crimped. In the initial stiff region micro-cracks start appearing as the composite specimen undergoes traction but the interface at resin and multifilament tows is still intact. That is why the composite exhibits rigid behaviour. In Fig. 10 it can be observed that after the initial stiff region the highly crimped tows tend to straighten due to increasing tensile load in the second region. In this region the micro-cracks give way to relative slippage of highly crimped tows in their sockets i.e., the resin-tow interface is relatively weakened. It can also be remarked that the sensor resistance follows the stress strain curve, but in the second region the electrical resistance curve is noisier as compared to other regions of the curve which might signify the slippage of tows as well as the sensor in their sockets. This second region is followed by the third region called the second stiff region where the tows are locked in their sockets. In this region the tows resist the applied load and exhibit stiff behaviour as they regain some of their initial stiffness after the straightening of tows in the second region. The electrical resistance varies almost linearly with the applied load, in this region. The third region is followed by the zone of rupture of the composite in which the electrical resistance, having attained the highest value starts dropping down. The normalized resistance starts dropping after the rupture. The fact that the sensor resistance attains its initial value after the rupture signifies that the sensor, owing to its elastic properties, is not destroyed with the composite. This fact was confirmed by tomographical image of the samples which underwent traction, shown in Fig. 11-a) and b). Sensor cross section and its path at and near the zone of rupture can be observed.

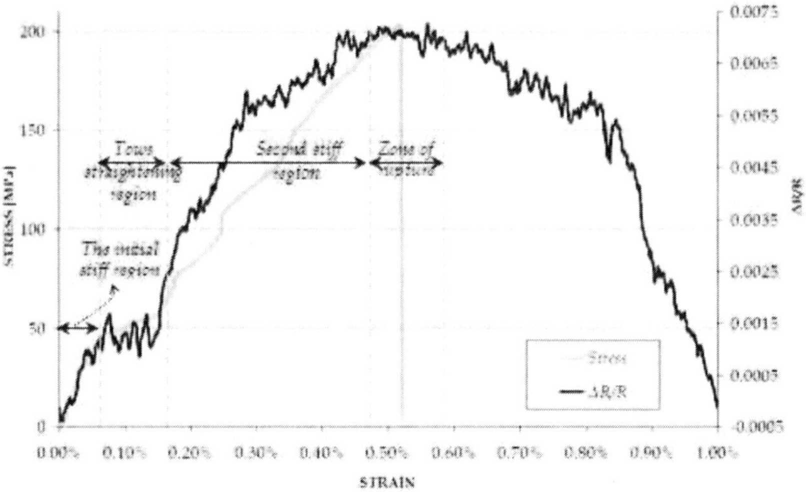

Figure 10. Normalized resistance and stress against strain for sensor inside composite

In Fig. 11-a) and b) it can be observed that the sensor-resin interface has a lot of voids. These are caused by poor resin-sensor interfacial properties. The insulating medium on the sensor surface needs to have good adherence with the epoxy resin and carbon fibre reinforcements. Damage that occurred at the main rupture zone has propagated along the sensor boundary giving rise to de-bonding of the sensor. A kink in the sensor can be observed which is caused by the relaxation of sensor as it tries to regain its original dimensions after the tensile loading damages the composite sample. The insulation coating around the sensor renders it thick as well which is undesirable for high performance composite materials as thick insulation coatings might adversely affect the mechanical properties.

3. HEATING ELEMENTS BASED ON CONDUCTIVE POLYMER COMPOSITE

The second part of this chapter presents a woven fabric containing an original heating element. Textile actuators like heating fabrics can find applications in numerous and varied fields such as sports, leisure, medical and automotive (Droval et al., 2005; El-Tantawy et al., 2002). In garments, wearability is affected because of the use of metallic components (heating wire and/or heating track on polymer flexible substrate), that are rarely elastic, flexible and lightweight. Nevertheless, these metallic, non-textile elements can be replaced by other conductive fibres such as silver plated polyamide fibres. In that case, the heating textile becomes lightweight, but very expensive (WarmX GmbH). In all the cases, heating systems need heavy power supplies. Thus, it is very important to develop heating textile systems able to work at low voltage.

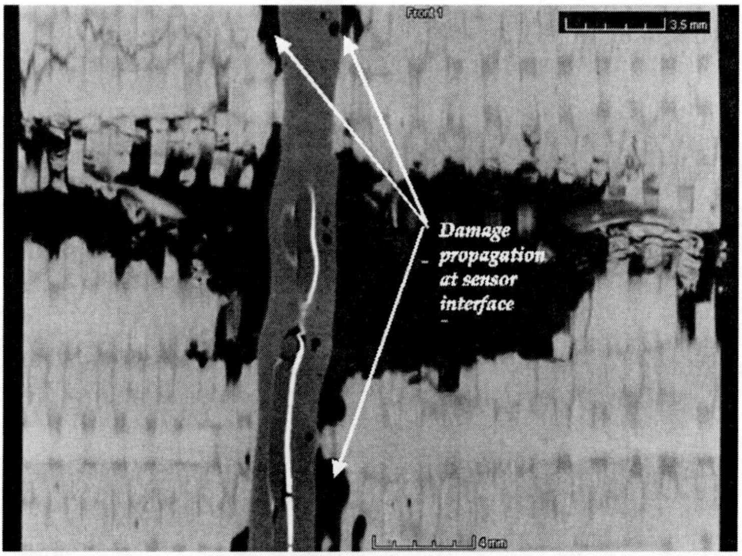

a) Frontal view

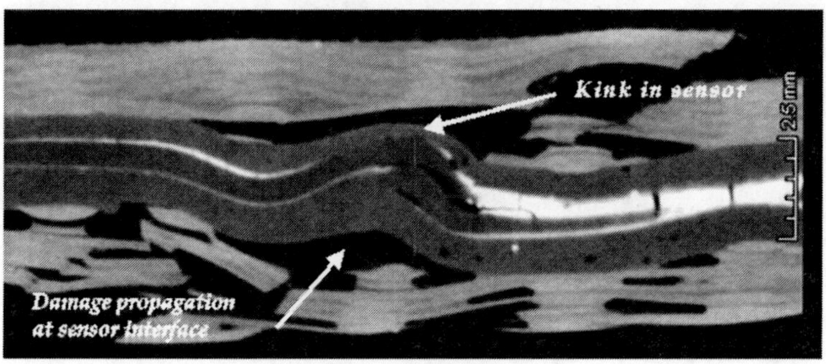

b) Longitudinal section

Figure 11. Tomographical images of sensor inside a tested sample near the zone of rupture

Our heating element is designed to adapt to woven flexible structures. Additional metallic yarns, used as electrodes, are integrated in a woven structure (or sewn into textile) in a comb-teeth arrangement. Function of these electrodes is to connect heating textile to a power supply and to distribute the current in the conductive coating layer applied on the fabric surface. The comb-teeth electrode arrangement is specially designed to ensure uniform heat distribution. The coating is realized with a composite material based on aqueous latex dispersed with carbon black (CB) as filler. The heating element (comb electrodes and electro-conductive coating) can thus adopt the desired pattern. This is an important aspect of our heating element as it allows integration of the heating element in various fabrics designed for varied and diverse applications.

3.1. Materials and Methods

Comb structure was made with stainless steel yarns (2 x 275 x 12 µm from Bekintex®). The average yarn count was 500 Tex, with a resistivity of 14 ohm/m. These yarns were either woven or sewn on an existing fabric. The common feature of all the configurations is that only one comb-teeth structure was used (Fig. 12). The textile fabric was woven on a hand loom (ARM loom equipped with Selectron command box). A plain weave was chosen. Cotton yarns were used in warp and weft having densities of 27 and 10 yarns/cm respectively. The stainless steel yarns were introduced manually during the weaving process according to the pattern (Fig. 12).

Samples with heating surface (*i.e.* L x l in Fig. 12) larger than 180 cm^2 were prepared. In typical samples, the dimension *L* was about 140 mm while *l* was about 150 mm. In this study, the distance between electrodes (*lp*) remained unchanged: i.e., 20 mm.

The coating was made using a conductive polymer composite (CPC) composed of carbon black (CB, Printex ® L6, Degussa), a synthetic rubber latex solution (Kraton® IR-401, Kraton Polymers) a dispersing agent (Disperbyk®-2010, SPCI) and water.

The preparation procedure is as follows: the dispersing agent is put into water and the CB particles are gradually added while mixing continuously. The polymer is finally added while mixing gently in order to avoid too strong shearing. The coating was then applied on the fabric with a magnetic coating table equipped with a magnetic bar as scraper. 12 samples were prepared with different CB content: 2.5, 5.0, 7.5, 10.0, 15.0, 20.0, 30.0, 35.0, 40.0, 45.0, 50.0 and 60.0 wt.-%. These contents were calculated from the total weight CB + Latex solution. After coating the woven fabric samples were dried at 50 °C for 12 hours. For all the samples, the thickness of the final coating layer was 450 ± 50 µm.

For each coating surface resistivity was measured using four-point probe (MR-1 Surface resistance meter, Schuetz Messtechnik). The aim of these measurements is to determine the percolation threshold and the minimum CB content which allows sufficient electrical conduction for our application.

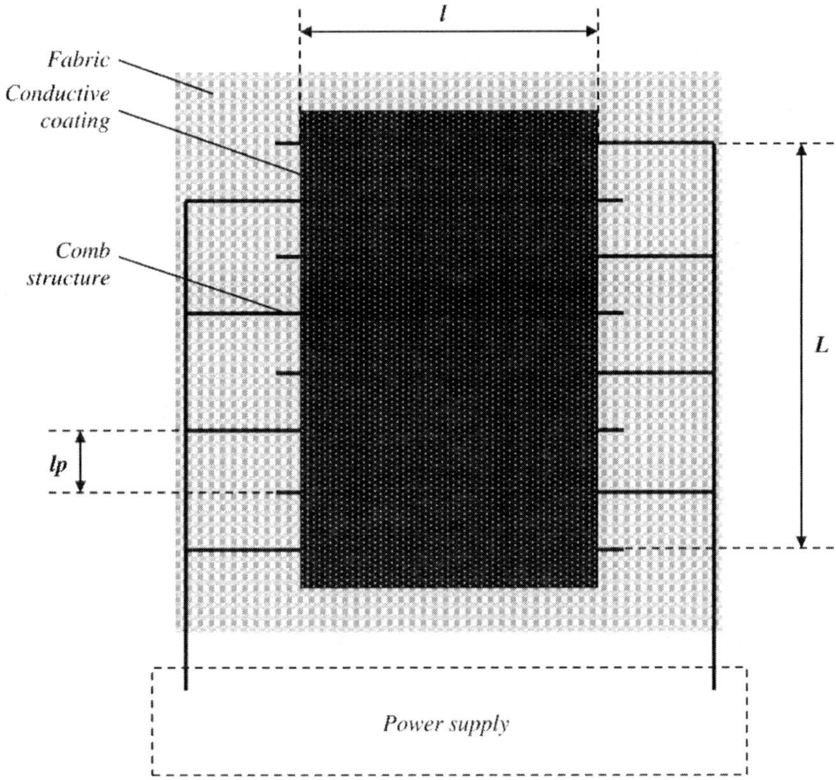

Figure 12. General structure (comb-teeth pattern and conductive coating) of heating textile element

To characterize heating effect of the samples, 2 processes were used:

- Feeding of the heating element with variable voltage supply (10, 15, 20 and 24 V). The surface temperature was recorded using thermocouple at 15 minute intervals at 5 different locations of the fabric. The average temperature was calculated from 5 measurements. Ammeter was used to determine the power consumption (W) of the heating element. This consumption is expressed in mW/cm^2 (taking into consideration the surface area of each sample),

- Feeding of the heating element with constant voltage supply (15 V) in conjunction with an IR camera (Agema ThermoVision 900). This camera took an IR image every 20 seconds.

3.2. Results

Fig. 13 shows electrical resistivity of coatings plotted against filler (CB) content in the latex solution. As expected, it is possible to identify the percolation threshold from this plot, which lies at 12 ±1 wt.-%. The form of the plot is in accordance with the typical behaviour of systems consisting of percolation networks (Kirkpatrick, 1973).

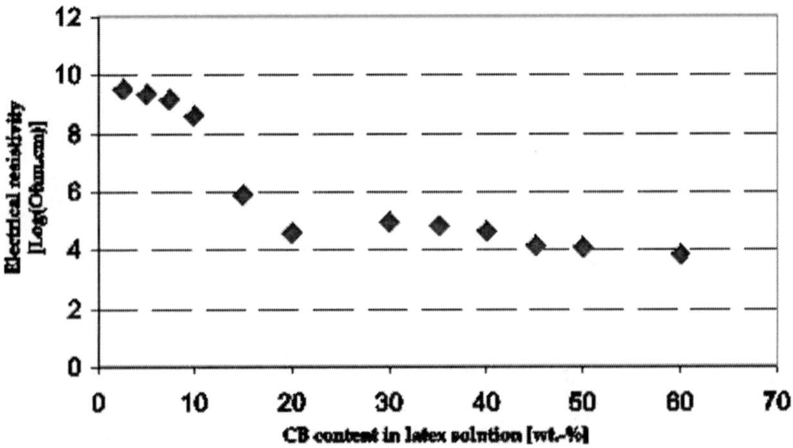

Figure 13. Electrical resistivity of the coating vs. CB content in the latex solution

This threshold value in wt.-% is expressed for liquid latex solution. Liquid latex contains approximately 63 % of dry material by weight. Thus, the corrected value of percolation threshold is near 18 wt.-%. This value is relatively higher than the value reported in literature for similar systems, (Grunlan et al. 1999, 2001). In our study, process of dispersion (including rupture of CB aggregate) and coating on fabric is not yet optimized. Obtained results show that 15 wt.-% of CB is necessary to obtain à conductive coating. Nevertheless, Fig. 13 shows that between 15 and 40 wt.-% resistivity is not optimal: therefore it is necessary to fill the composite at least by 45 wt.-% to obtain lower resistivity.

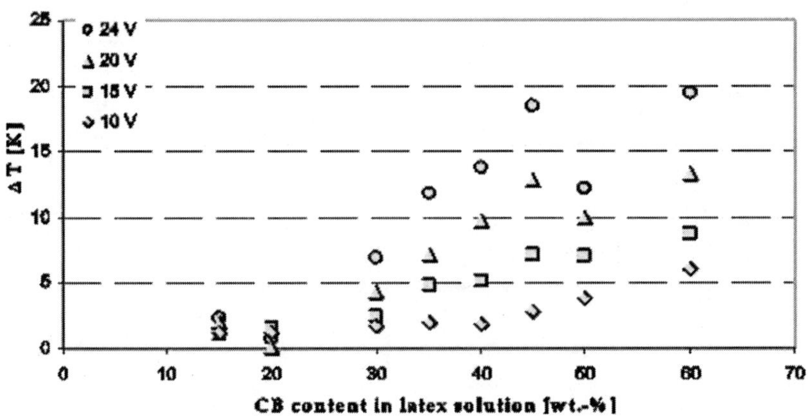

Figure 14. Surface temperature vs. CB content for feeding voltage of 10, 15, 20 and 24 V

Fig. 14 shows surface temperature of coating plotted against the filler content (in liquid latex) for several feed voltages (10, 15, 20 and 24 V). Temperature on the graph (ΔT) is expressed as difference between measured temperature and room temperature (between 20 and 22 °C). No elevation of temperature was recorded for sample under 30 wt.-% of CB. For CB content between 30 wt.-% and 45 wt.-%, ΔT increases with the CB content. Above 45 wt.-% of CB, ΔT does not increase significantly with filler addition. These results are in agreement with the previous remarks concerning resistivity vs. CB content.

These results show that the best content of CB was, in our case, 45 wt.-%. Under this value heating effect was non optimal. Above this value addition of CB does not increase heating ability.

Maximum value of ΔT (near 20 K) was registered for sample voltages of 24 V and CB content of 45 wt.-% and 60 wt.-%. Fig. 15 shows that for these heating elements, electrical input power was close to 250 mW/cm².

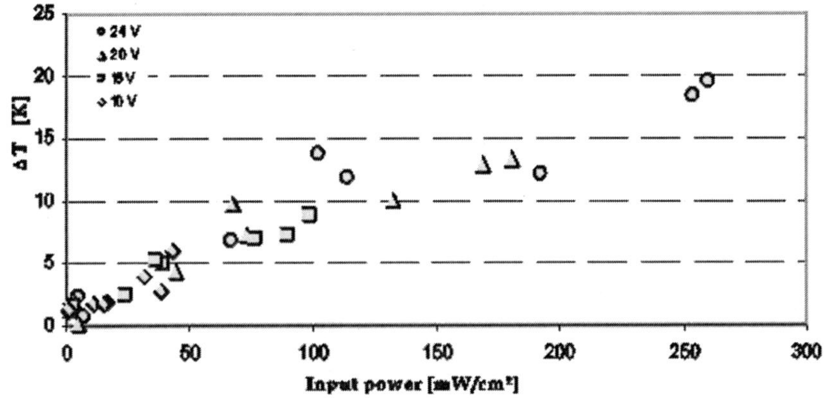

Figure 15. Increase of surface temperature (ΔT) vs. power consumption of heating element

Infrared images show the distribution of heat in the structure (comb electrodes and conductive coating) vs. time. Fig. 16 shows temperature of sample with 60 wt.-% of CB from t = 0 s (Fig. 16-a) to t = 180 s (Fig. 16-f). Feeding voltage was constant and was equal to 15 V.

Fig. 16 shows that comb structure (stainless steel yarn) heats first after switching on. The maximum temperature of this yarn was about 70 °C. This temperature is achieved after 120 s. CB coating heats relatively slowly but it can be conjectured that after 120 s the temperature of all the surface area exceeds 35 °C while it exceeds 40 °C after 180s. This behaviour is expected since stainless steel yarns have better thermal conductivity than carbon based composite.

4. CONCLUSION

The sensor based on conductive polymer composite, developed for in situ measurements on carbon fibre based woven fabric composite, is capable of detecting strain in the structure. The electrical resistance variation in the sensor follows the deformation pattern of the composite, mainly due to its sensitivity to its environment and because of the fact that it is integrated in the structure and follows the fibre architecture of the reinforcement. It has been shown that the integrated textile sensors inside the reinforcement can be used as in situ strain gages for the composite materials. Moreover, if the placement of these sensors inside the reinforcement is carefully chosen, they can be used to follow the local deformation pattern so as to better understand the deformation mechanisms and predict life time of the composite parts. At present the sensors have been tested for tensile loading. Tensile strength tests were chosen to demonstrate the basic features of this novel SHM approach. In the future these sensors will be used for bending and fatigue tests on similar 3D carbon fibre woven reinforcement based composites. However optimisation of sensors needs to carried out in order to prepare finer sensors having negligible affect on reinforcement geometrical and mechanical properties. For carbon fibre based reinforcements which require an insulation coating on the sensor surface, a better and finer coating needs to be applied. In view of the test results presented, it can be concluded that these sensors can be used for in situ health monitoring of various types of composites for industrial applications (aeronautics, automotive etc).

We have also developed a heating element based on original comb-teeth structure (stainless steel electrode) and electro-conductive coating composed of latex and carbon black. Comb structure can be either woven or sewn into the fabric. Final product is flexible and lightweight. This study has show that ideal carbon content of the coating was 45 wt.-% in a latex solution. Our heating elements (about 200 cm^2) allow increasing the temperature (ΔT) by 20 K with low voltages (between 15 and 24 V). Temperature homogeneity of heating elements is better than those of heating elements made with only stainless steel yarns because in the later case, heat is only produced and localized at conductive yarns. Moreover, our system is cost efficient because we used a few stainless steel yarns.

The next step of this study would allow, in the first place, the optimization of composite preparation (liquid Latex + CB) and secondly the realization of larger heating elements (> 0.5 m^2). Potential applications of these heating elements can be found in garments (to improve thermal comfort), in transportation where heating is required for passenger comfort and in certain industrial systems (antifreeze).

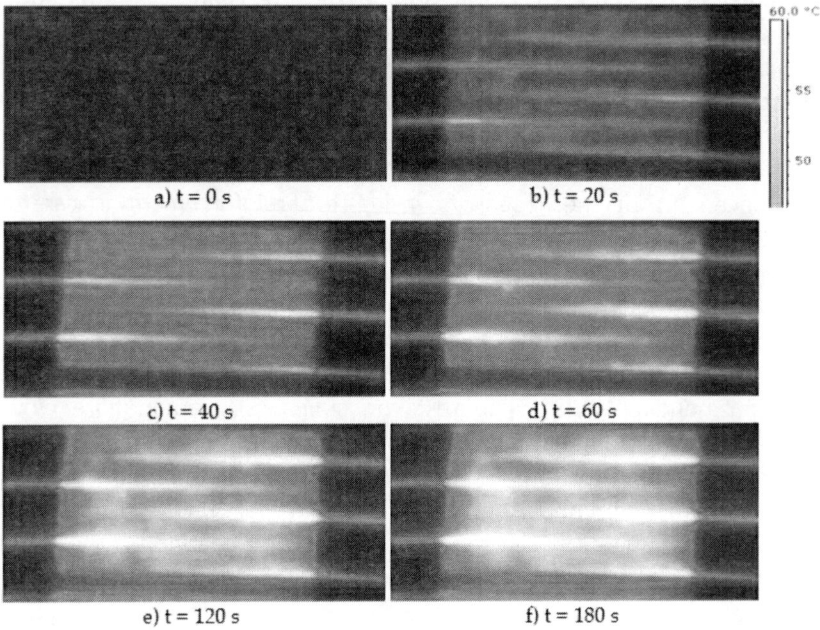

Figure 16. IR image of 60 wt.-% CB heating element from a) t = 0 s to f) t = 180 s, voltage = 15V

ACKNOWLEDGEMENTS

This research was partially financed by Interreg IV via TRITEX (Transfer of Research and Innovations in TEXtiles) program (FLV1.1.1). Special acknowledgments to Textile Department Laboratory from UGENT (Belgium) for electrical measurements (part 3.) and use of IR camera.

REFERENCES

1. S. Black, 2009 Structural health monitoring: Composites get smart, Composites World, http://www.compositesworld.com/articles/structural-health-monitoring-composites-get-smart

2. C. Cochrane, V. Koncar, M. Lewandowski, C. Dufour, 2007 Design and development of a flexible strain sensor for textile structures based on a conductive polymer composite. Sensors, 7 473492.

3. C. Cochrane, M. Lewandowski, V. Koncar, 2010 A Flexible Strain Sensor Based on a Conductive Polymer Composite for in situ Measurement of Parachute Canopy Deformation. Sensors, 10 82918303.

4. P. Dharap, L. Zhiling, S. Nagarajaiah, E. V. Barrera, 2004 Nanotube film based on single-wall carbon nanotubes for strain sensing, Nanotechnology, 15 379382.

5. G. Droval, P. Glouannec, J. F. Feller, P. Salagnac, 2005 Simulation of electrical and thermal behavior of conductive polymer composites heating elements. Journal of Thermophysics, 19 3 375381.

6. F. El -Tantawy, K. Kamada, H. Ohnabe, 2002 In situ network structure, electrical and thermal properties of conductive epoxy resin-carbon black composites for electrical heater applications. Materials Letters, 56 112126.

7. European Standard NF EN ISO5274 1997 (), Plastiques- Détermination des propriétés en traction, Partie 4: Conditions d´essai pour les composites plastiques renforcés de fibres isotropes et orthotropes.

8. B. Fiedler, F. H. Gojny, M. H. G. Wichmann, W. Bauhofer, K. Schulte, 2004 Can carbon nanotubes be used to sense damage in composites?, Annales de chimie, 29, 8194.

9. C. L. Gettinger, A. J. Heeger, D. J. Pine, Y. Cao, 1995 Solution characterization of surfactant solubilized polyaniline. Synth. Met., 74, 8188.

10. J. Grunlan, W. Gerberich, L. Francis, 2001 Lowering the percolation threshold of conductive composites using particulate polymer microstructure. Journal of applied polymer science, 80 4 692705.

11. J. Grunlan, W. Gerberich, L. Francis, 1999 Electrical and mechanical property transitions in carbon-filled poly(vinylpyrrolidone). Journal of materials research, 14 11 41324135.

12. Y. Haba, E. Segal, M. Narkis, G. I. Titelman, A. Siegmann, 1999 Polymerization of aniline in the presence of DBSA in an aqueous dispersion. Synth. Met., 106 5966.

13. A. J. Heeger, 2002 Semiconducting and metallic polymers: the fourth generation of polymeric materials. Synth. Met. 125 2342.

14. R. Kamiya, B. A. Cheeseman, P. Popper, T. Chou, W., 2000 Some recent advances in the fabrication and design of three-dimensional textile preforms: a review. Composites Science and Technology, 60 3347.

15. S. Kirkpatrick, 1973 Percolation and conduction. Reviews of Modern Physics, 45 4 574588.

16. F. Ko, 2007 3 -D textile reinforcements in composite materials, 3-D Textile Reinforcements In Composite Materials. vol. null: Woodhead Publishing Limited.

17. I. Krupa, I. Chodak, 2001 Physical properties of thermoplastic/graphite composites. Europ Polym J., 37, 21592168.

18. D. Kumar, R. C. Sharma, 1998 Advances in conductive polymers. Eur. Polym. J. 34 10531060.

19. C. Li, E. T. Thostenson, T. W. Chou, 2008 Sensors and actuators based on carbon nanotubes and their composites: A review, Composites Science and Technology, 68 12271249.

20. F. Lorussi, E. P. Scilingo, M. Tesconi, A. Tognetti, D. De Dossi, 2005 Strain sensing fabric for hand posture and gesture monitoring, IEEE Transactions On Information Technology In Biomedicine, 9 372381.

21. I. Novak, I. Krupa, I. Chodak, 2002 Investigation of the correlation between electrical conductivity and elongation at break in polyurethane-based adhesives. Synth. Met., 131 9398.

22. E. P. Scilingo, F. Lorussi, A. Mazzoldi, D. De Rossi, 2003 Strain-sensing fabrics for wearable kinaesthetic-like systems, IEEE sensors journal, 3 460467.

23. J. Wilson, 2004 Sensor signal conditioning, Sensor Technology Handbook. vol. null: Newnes Publishing Limited.

24. www.nottingham.ac.uk/~emxmns/texgen.htm, consulted on 10/01/2011

25. www.datasheetdir.com/INA101 -Instrumentation-Amplifiers, consulted on 06/05/2010.

Chapter 3

Influence of fibre architecture on the tensile, compressive and flexural behaviour of 3D woven composites

S. Dai[a], P.R. Cunningham[a], S. Marshall[b], C. Silva[b]

[a]Department of Aeronautical and Automotive Engineering, School of Aeronautical, Automotive, Chemical and Materials Engineering, Loughborough University, LE11 3TU, UK

[b]M. Wright and Sons Ltd, Quorn Mills, Quorn, Loughborough LE12 8FZ, UK

ABSTRACT

This paper presents a comprehensive study on the tensile, compressive, and flexural performance of six types of 3D woven carbon-fibre/epoxy composites which were manufactured using a traditional narrow fabric weaving loom and resin transfer moulding. Four orthogonal and two angle-interlock weaves were tested with the primary loading direction parallel to the warp direction. The mechanical performance was found to be affected by the distribution of resin rich regions and the waviness of the load-carrying fibres, which were determined by the fibre architectures. The binding points within the resin rich regions were found to be the damage initiation sites in all weave types under all loading conditions, which were confirmed with both visual observation and digital image correlation strain maps. Among all weave types, the angle interlock weave W-3 exhibited the highest properties under all loading conditions.

KEYWORDS

A. 3-Dimensional reinforcement; B. Mechanical properties; E. Weaving; E. Resin transfer moulding (RTM)

1. INTRODUCTION

The current trend in composites research in the aerospace and automotive industry is to develop advanced composites using low cost "out-of-autoclave" manufacturing techniques, produce a component with high structural integrity and high delamination resistance, and explore the potential for automated manufacturing processes [1],[2] and [3]. Compared to conventional pre-preg layered laminated composites, 3D woven composites provide these advantages. It has been shown by various researchers that the impact resistance, post-impact strength, and delamination toughness are improved by introducing through-thickness binder fibres [1], [4], [5], [6] and [7]. However, the in-plane mechanical properties of 3D woven composites are generally compromised due to the existence of fibre crimping introduced by the weaving process [1] and [8]. Three-dimensional woven composites have shown both increases and decreases in elastic modulus, tensile strength, and compressive strength compared with the conventional 2D laminated composites which have comparable in-plane fibre structures (such as similar in-plane fibre volume fraction) [2]. However, due to the limited published data for flexural strength, it is difficult to determine whether the flexural strength is increased or decreased by the through-thickness fibres [2]. Comprehensive experimental studies were carried out on the tensile, compressive, and flexural behaviour of 3D woven composites [7], [9], [10],[11], [12] and [13], however more data are required to further understand the behaviour of 3D woven composites with various weave architectures and update with the advances in weaving technology.

The in-plane tensile properties of various types of 3D woven composites have been experimentally characterised and compared with other fibre architectures such as non-crimp fabrics and 2D woven composites [8], [9], [10], [11], [14], [15], [16] and [17]. One of the common conclusions drawn by this research is that the tensile properties can be improved by having a minimum waviness in the in-plane fibres, which is affected by the weave architecture. The compressive properties of 3D woven composites were reported to be lower [18], similar [12], and even higher [7] compared to the 2D equivalent composites. It was suggested by Cox et al. [19] that it was more efficient to improve the compressive strength by increasing the geometrical regularity than increasing the fibre volume fraction, which on the other hand results in brittle compressive failure and lower failure strain. The compressive failure was found to initiate around the geometrical flaws and was dominated by fibre kinking which almost always occurred across the entire cross-section of the load-carrying tows [11], [19], [20] and [21]. The effect of the weave architectures on the flexural behaviour of 3D woven composites were studied in [9],[12] and [22], and the through-thickness binder yarns were found to be effective in resisting delamination crack growth. Kuo [23] conducted flexural tests on two types of 3D orthogonal woven carbon/epoxy composites and concluded that the binder yarn loops at the surface of the fabrics prevent and deflect crack propagation. Adanur and Tam [24]reported that the flexural strength of 3D interlock glass/epoxy composites was higher than 2D laminates.

In this paper, four types of 3D orthogonal woven composites and two types of 3D angle interlock woven composites were manufactured and tested under tension, compression, and three-point bending. The manufacturing process and experimental techniques are described, and the influence of the weaving process on the weave architectures are discussed. The stress–strain and load–displacement curves under different loading conditions are presented along with the tensile, compressive, and flexural properties, and full-field surface strain distribution maps during tensile tests are illustrated using a digital image correlation system. The results of all six weave architectures are compared and discussed.

2. MATERIALS AND MANUFACTURE TECHNIQUE

2.1. Manufacturing process

All of the fabric preforms were woven using a traditional narrow fabric weave loom (Muller-NC2-S) by M.Wright & Sons Ltd. Six types of fabrics were produced with the geometric model of the idealised weave architectures generated using TexGen [25] as shown in Fig. 1. One design of a 1-by-1 orthogonal weave (W-1), three designs of a 3-by-3 orthogonal weave (W-2.1, W-2.2, and W-2.3), and two types of angle-interlock weaves (W-3 and W-4) were produced. All six weaves used the same type of warp and weft yarns as listed in Table 1, and the binder yarns were different between designs in order to achieve the same amount of fibres in the through-thickness direction. The spacing of the warp tows was kept at 1.535 ends/mm and the weft tow spacing was kept at 1.5 picks/mm for all of the weaves. The nominal thickness was 0.42 mm for the weft tow and 0.40 mm for the warp tow, and the nominal width was 1.7 mm for the warp tow and 2.5 mm for the weft tow. It should be noted that each weft tow consists of two 6k yarns because the loom employed a rapier weft insertion system which feeds one weft yarn during weft insertion and another weft yarn during weft arm retraction.

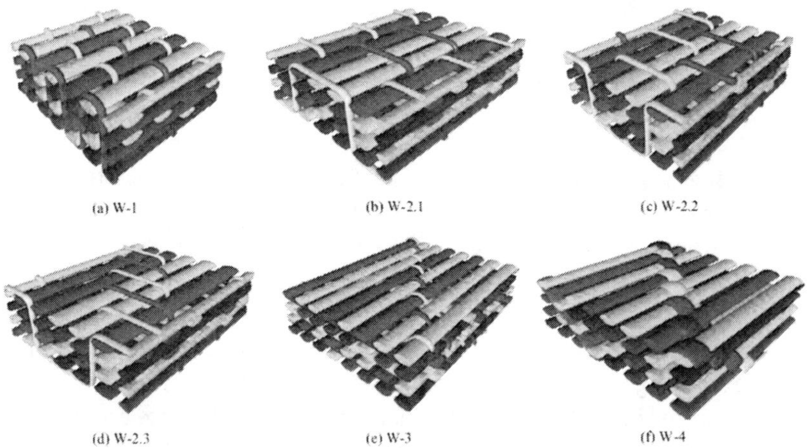

Figure 1. Idealised weave architectures generated using TexGen. (For interpretation of the references to colour in this figure legend, the reader is referred to the web version of this article.)

Each preform was an 80 mm wide, 350 mm long, and 3 mm thick (all dimensions are nominal) strip since it was fabricated using a traditional narrow fabrics weaving loom. Five strips were placed in a rectangular shaped closed mould tool for resin transfer moulding using a Hypaject MK-III RTM system. An 8 mm silicone intensifier was use to fill in the redundant cavity in the mould tool and provide a consolidation pressure. A Gurit Prime 20LV epoxy resin system, which is a two parts epoxy system consisting of resin and slow hardener, was used to infuse the preforms. The resin was degassed and heated up to 30 °C in the homogeniser and then injected into the vacuumed and preheated mould tool. The injection pressure was kept at 1 bar before the mould tool was fully filled, and once the resin filled the mould the outlets were locked and the injection pressure was increased to 1.5 bar and kept for 5 min to further fill any possible dry spots. The mould tool was then heated up to 50 °C for 16 h for curing. The entire manufacturing process including both weaving and moulding were carefully documented and monitored to ensure consistency of the samples.

2.2. Examination of the manufactured weave structures

The manufactured samples were sectioned, polished and examined under an optical microscope. All microscopic samples were randomly selected from the same panel from which the testing samples were cut, and the typical internal and surface fibre geometries of the six weaves are shown in Fig. 2 and Fig. 3. As can be seen, the actual fibre architectures differ from the idealised designs. For instance, the warp and weft tows were designed to be non-crimp in all of the designs, however the real samples showed a certain degree of waviness, which was quantified for the warp tows as shown in Table 4. Six warp tows were taken for measurement from the microscopic images to give the averaged tow waviness. A more detailed discussion of the weave structures is presented in the following sections.

2.2.1. Binder/warp direction

The actual binder path of the weaves are affected by the interlacing movement which largely depends on the weave architectures. W-1 had the most orthogonally placed binder tows among all four weaves. The 1-by-1 orthogonal weave had a more compact binding structure than the other 3-by-3 weaves since every column of the weft tows were interlaced by the binder yarns in W-1. This compact interlacing sequence resulted in smaller resin rich regions and less free space for the warp tows to move and therefore relatively straight warp tows.

Table 1. Preform specification.

Weave type	ID	Binder tow		Warp tow		Weft tow	
		Type	Counts	Type	Counts	Type	Counts
Orthogonal 1 × 1	W-1	Toray T300	1k	IMS5131	24k	HTA40E13	2 × 6k
Orthogonal 3 × 3	W-2.1	Tairyfil T33	3k	IMS5131	24k	HTA40E13	2 × 6k
	W-2.2	Tairyfil T33	3k	IMS5131	24k	HTA40E13	2 × 6k
	W-2.3	Tairyfil T33	3k	IMS5131	24k	HTA40E13	2 × 6k
Angle interlock	W-3	Tairyfil T33	3k	IMS5131	24k	HTA40E13	2 × 6k
	W-4	IMS5131	24k	N/A	N/A	HTA40E13	2 × 6k

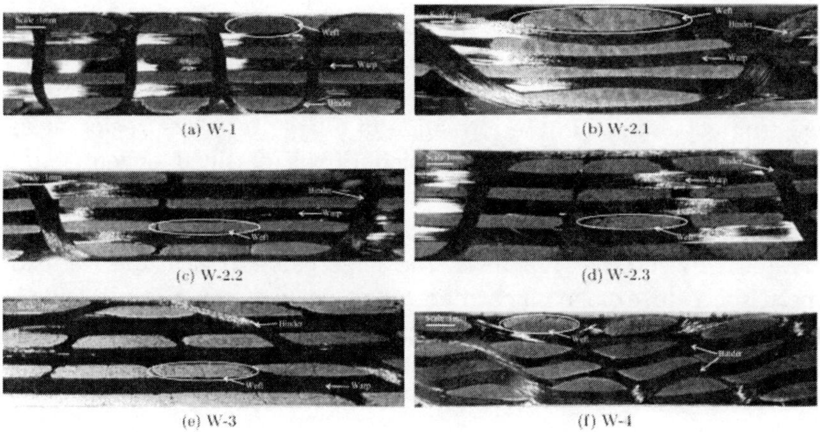

Figure 2. Internal geometry of woven composites, section parallel to binder direction. (For interpretation of the references to colour in this figure legend, the reader is referred to the web version of this article.)

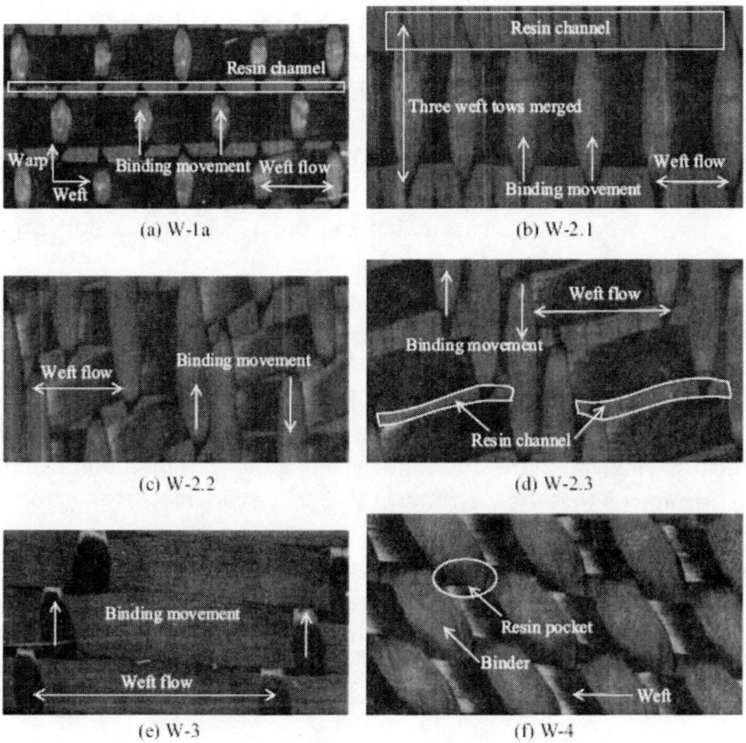

Figure 3. Surface geometry of woven composites.

In W-2.1, three weft tows were bound together, which resulted in the merging of these three weft tows forced by the binding movement. Therefore the space designed to be in between two individual weft tows was shifted resulting in the formation of a large resin rich region at each side of the merged tows. Since the binder tows remained tangential to the weft tows at the surface of the weave, the through-thickness portions of the binder tows were then inclined at an angle within the extra space created between the weft tows. This large gap also gave the warp tows more space to expand since there were no weft tows constraining the movement. It is clearly shown in Fig. 2 that the warp tow of W-2.1 had a wavy path and expands its cross-section in the through-thickness direction between the two merged weft tows.

Weave W-2.2 and W-2.3 were also 3-by-3 weaves similar to W-2.1 with three weft tows bound together by one binder yarn but in different binding sequences. However, these three weft tows were separated by another binder in the adjacent cell, which kept the weft tow within the vicinity of its designed position. Hence, the weft tow spacing was more evenly distributed in these two weaves, which resulted in smaller resin rich regions and more orthogonally placed binders compared to W-2.1. Moreover, the warp tows were relatively straight and had a more constant cross-section along the length in these two weaves than in W-2.1. The warp tows in W-2.2 were slightly less crimped than in W-2.3 due to a shorter weft flow as shown in Fig. 4. Within the longer weft flow, the weft tows were relaxed and moved towards an adjacent binding point, which then left a gap in between that allowed the warp tow to crimp.

Weave W-3 is an angle-interlock weave with non-interlacing warp tows which had limited crimp. The cross-section of the weft tows in W-3 shifted into a parallelogram-like shape to fit the binder tow path. Since two adjacent binder tows had different paths, the parallel angle of the cross-section also changed in the adjacent cell along the weft direction. Therefore it can be inferred that the weft tow in W-3 had a varying cross-section.

Weave W-4 is an angle-interlock weave with only interlocking binder tows which used the same IMS5131-24k yarns as the warp tows in all other designs. As shown in Fig. 2, all binder tows were at an angle and the cross-sections of the weft tows were shaped into various forms by the tensioning force on the 24k binder yarns during the weaving process.

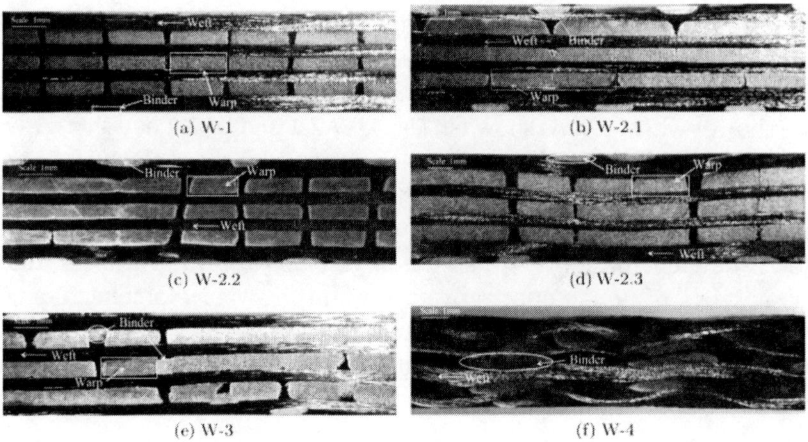

Figure 4. Internal geometry of woven composites, section parallel to weft direction.

2.2.2. Weft direction

In all of the orthogonal weaves, due to the orthogonal binder insertion, a clear resin channel was left between two weft tows as shown in Fig. 3. It can also be seen that W-1 had a evenly spaced resin-rich channel, W-2.1 had the widest resin rich channel, and both W-2.2 and W-2.3 had discontinuous and curved resin channels. The effect of the resin channel on the mechanical performance will be discussed in a later section. In the angle-interlock W-3, the surface resin rich regions were localised around the binding point and did not form a distinctive channel. In W-4, resin rich zones were distributed as "resin pockets" between the weft and binder tows.

The in-plane waviness of the weft tows varies between designs. As shown in Fig. 4, the weft tows of W-1, W-2.1, W-3, and W-4 had the least in-plane waviness, because the adjacent binders on these weft tows had the same binding movement which did not force the weft tows to bend in opposite directions. While in W-2.2 and W-2.3, the two adjacent binders had the opposite movement as demonstrated by the arrows in Fig. 3. The curvature of the weft tows at the binding points in W-2.2 was larger than in W-2.3, which was caused by the difference in binding sequence between the two weaves.

The samples were also sectioned along the weft direction to reveal the internal weft tow path, as shown in Fig. 4. W-1 had the least crimped weft tows among all of the designs due to the compact binding sequence and the shorter weft flow. As shown in Fig. 4, W-2.1 and W-2.2 had shorter weft flows than W-2.3, therefore the out-of-plane waviness of the weft tows in these two weaves was larger than in W-2.3. W-3 also had some extent of out-of-plane waviness in the weft tows, which is caused by the longer weft flow. The weft tows in W-4 had the largest out-of-plane waviness due to the absence of non-interlacing warp tows in this weave architecture. The non-interlacing 24k warp tows in all other designs were changed into angle-interlock tows in W-4. This increased thickness angle-interlock binder yarn also caused the out-of-plane waviness of the weft.

2.3. Measured fibre volume fraction

The nominal thickness of the infused composite panel was 2.78 mm with variations between weave types as listed in Table 2, which were averaged from the 22 mechanical testing samples. The fibre volume fractions of the six types of composites were obtained from matrix burn-off tests. By measuring the weight before and after burn off tests, the overall fibre volume fraction were calculated. The fibre volume fraction of the warp tows and weft tows were measured by separating a dry fabric strip which was long enough to cover at least four complete unit cells, into individual tows and weighing the overall mass of each type of tow. Two dry fabric samples and three composite samples were measured and the averaged results are listed in Table 2. All of the weave architectures except W-3 had a similar overall fibre volume fraction of around 50%, while W-3 had the lowest overall fibre volume fraction and warp fibre volume fraction.

3. MECHANICAL TESTING

All samples were cut from the moulded panel using a water jet cutter. Tensile, compression, and flexure tests were conducted on all six weave architectures with the primary loading axis parallel to the warp direction. Five samples were tested for each weave type under each loading condition and the geometries of each type of samples are shown in Table 3.

Table 2. Fibre volume fraction and thickness of the samples.

ID	V_f (%)	CV (%)	V_{warp} (%)	CV (%)	V_{weft} (%)	CV (%)	t (mm)	CV (%)
W-1	49.86	1.59	27.72	4.59	21.10	5.83	2.71	3.48
W-2.1	49.86	3.46	29.98	5.43	17.73	9.64	2.85	5.81
W-2.2	50.02	4.11	27.84	1.79	19.57	3.40	2.77	5.47
W-2.3	51.16	3.32	28.20	1.38	20.55	2.43	2.80	5.00
W-3	46.01	2.57	25.66	3.02	18.09	3.18	2.77	6.11
W-4	50.98	2.35	N/A	N/A	21.42	4.35	2.72	2.62

Table 3. Nominal sample geometries.

Test	Width (mm)	Length (mm)	Gauge (mm)	Shape
Tensile	25	250	150	Rectangular
Compressive modulus	15	80	10	Rectangular
Compressive strength	15	80	5	End-tabbed rectangular
Flexure	25	160	120	Rectangular

3.1. Tensile testing

The tensile tests were carried out following ASTM D3039 standard [26] using an Instron 6025 testing machine with a 100 kN load cell as shown in Fig. 5. A displacement control of 2 mm/min was applied and one 5 mm long TML-BFLA-5-3 strain gauge was attached in the centre of each specimen along the loading direction to determine the elastic modulus. Since the strain gauge only measures local strain and fails at higher strain, a LaVision 2D digital image correlation (DIC) system was used on two extra samples of each weave in order to obtain full-field strain development especially in the higher strain regions. Random speckle patterns were applied to these samples using white paint on a matt black base coat, and the DIC camera recorded the deformation at 4 frames/s during loading. All of the samples were loaded until cross-sectional fracture with the load, displacement, and strain data recorded at 2 Hz.

3.2. Compression testing

The compression tests were conducted following the modified ASTM D695 standard[27] and [28] using an Instron 6025 with a 100 kN load cell and an anti-buckling guide as shown in Fig. 6. In order to achieve an acceptable compressive failure mode a short gauge length of 5 mm was required [27], which was not enough to attach the strain gauges. Therefore two independent tests were conducted to obtain the compressive modulus and strength. The compressive modulus were obtained from strain gauged un-tabbed samples, while the compressive strength were obtained from end-tabbed samples. For each untabbed sample, another sample was cut next to it to obtain the compressive strength. A displacement control of 2 mm/min was applied on each sample and TML-BFLA-2–3 (2 mm) strain gauges were attached in the centre of the un-tabbed specimens.

Figure 5. Experimental set-up for tensile tests with a DIC system. (For interpretation of the references to colour in this figure legend, the reader is referred to the web version of this article.)

Figure 6. Experimental set-up for compression modulus tests. (For interpretation of the references to colour in this figure legend, the reader is referred to the web version of this article.)

3.3. Flexure testing

The flexural properties were obtained from three point bending tests. Preliminary tests were carried out on W-1 samples with span-to-thickness ratio of 16, 32, 40, and 60 as suggested by ASTM D790 [29], and five samples were tested with each ratio. It was found that a span-to-thickness ratio of 40:1 provided the least scatter in the flexural modulus and strength results, which was then used to conduct all of the flexural tests. An Instron 8870 with a 25 kN load cell was used to perform the tests as shown in Fig. 7, and a 2 mm/min displacement control was applied.

Figure 7. Experimental set-up for flexure tests. (For interpretation of the references to colour in this figure legend, the reader is referred to the web version of this article.)

4. RESULTS AND DISCUSSION

4.1. Tensile testing

Fig. 8 shows the stress–strain curves of one example specimen from each weave with the stress and strain recorded until failure, and Table 4 lists the averaged tensile strength and modulus from all seven specimens. As shown in the stress–strain curves, all six samples exhibited linear behaviour at the beginning of loading and non-linear behaviour towards final fracture. For all other five weaves except W-4, the non-linear behaviour was attributed to the damage occurring in the specimens and subsequent load re-distribution, such as matrix cracking. For W-4, the non-linear behaviour was partially caused by the damage and partially caused by the straightening effect of the angled binder tows. The linear portion of the stress–strain curve (0.1–0.3% strain) was used to calculate the tensile modulus according to ASTM D3039 [26]. Unfortunately the strain gauges on two thirds of the samples failed before the sample reached its tensile strength, possibly because the surface matrix cracks occurred within the gauged regions which resulted in the strain gauges debonding. Therefore failure strain was not obtained through these tests. Although the strain gauges failed, the initial linear portions of the stress–strain curves were still considered to be valid, because these linear regions were at about 0.6% strain (at least 200 data points) away from where the strain gauge started to fail. In addition, it is not uncommon to have strain gauges fail due to surface damage [21],[27] and [30]. As can be seen from Table 4, the tensile modulus was affected by the waviness of the warp tows: higher waviness resulted in lower modulus. Since the volume fraction of the load-carrying fibres plays an important role in the mechanical properties of composite materials, the tensile properties were normalised by the warp fibre volume of each weave fibre architecture in order to perform an appropriate comparison. The results were normalised according to Eq. (1) and are plotted in Fig. 9.

$$P_n = P_{act} \times \frac{28\%}{V_{warp}} \tag{1}$$

where P_n is the normalised material property, P_{act} is the actual material property, V_{warp} is the actual warp fibre volume fraction of each weave, 0.28 is the normalised warp fibre volume fraction averaged from all of the weaves.

Weave W-1 had the second highest normalised tensile strength and the third highest normalised tensile modulus. The compact weave structure of W-1 resulted in less crimped warp tows which led to the higher tensile properties. In addition, this compact pattern also improved the regularity of the weave structure, which resulted in W-1 having the lowest coefficient of variation. A similar weave pattern with different types of yarns was tested in [3] and showed a tensile modulus of 60 GPa and a tensile strength of 953 MPa in the warp direction. The warp yarns used in [3] had lower tensile properties than the warp yarns used in this study, and were also lower in volume fraction.

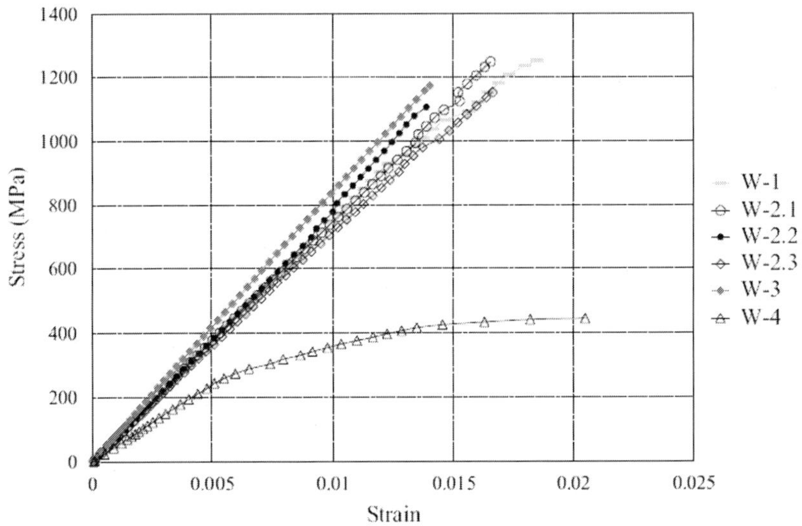

Figure 8. Typical stress–strain curves for all six types of woven composites under tension. (For interpretation of the references to colour in this figure legend, the reader is referred to the web version of this article.)

Table 4. Tensile properties.

Weave ID	Warp tow waviness (%)	Modulus (GPa)	CV (%)	Strength (MPa)	CV (%)
W-1	1.37	76.75	5.56	1358.54	3.22
W-2.1	2.83	69.44	29.53	1227.94	7.55
W-2.2	1.06	78.67	15.25	1103.17	6.61
W-2.3	1.40	73.46	10.48	1220.31	6.03
W-3	0.55	80.49	8.69	1276.24	5.96
W-4	N/A	48.54	13.44	409.90	4.91

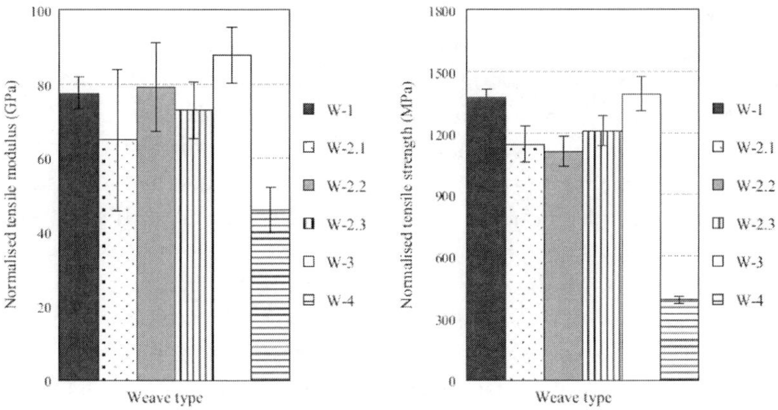

Figure 9. Normalised tensile properties.

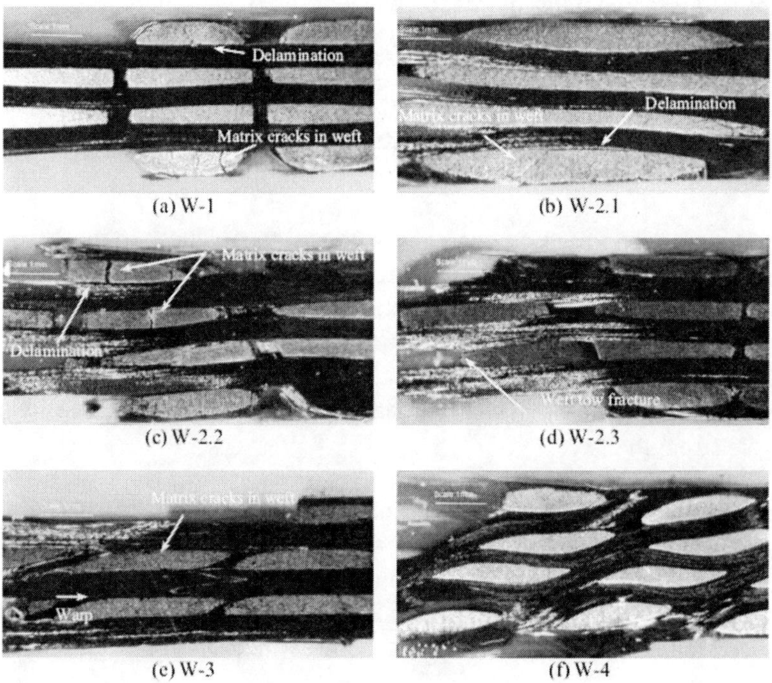

Figure 10. Failed sections of tensile samples (parallel to warp direction). (For interpretation of the references to colour in this figure legend, the reader is referred to the web version of this article.)

Weave W-2.1 showed the second lowest tensile modulus and the largest scatter. As discussed earlier, W-2.1 had a large resin rich region between the two merged weft tows and the warp tows had more waviness than other four weaves, which resulted in the lower tensile modulus. The unit cell of W-2.1 was relatively large and the fibre/resin distribution were relatively localised on the surface. Therefore the weave had localised high strain regions within the resin rich zones on the surface of the samples as shown in Fig. 11. The length of the strain gauge (5 mm) was not long enough to cover both the resin and fibre regions (10 mm) in this particular weave. Therefore the resultant modulus was artificially lower from the samples with the strain gauge that only covered the resin rich region, and artificially high modulus was obtained from the samples with the strain gauge covering the fibre region. This resulted in the larger coefficient of variation of W-2.1 and also explained the fact that the tensile modulus data had higher scatter than the tensile strength data. The large resin rich channels and crimped warp tows in W-2.1 were reduced by varying the binding sequence, as can be seen in W-2.2 and W-2.3, which resulted in higher tensile modulus of these two weaves. In addition, W-2.2 had a higher tensile modulus than W-2.3 due to its less crimped warp tow. However the tensile strength of W-2.2 was lower than W-2.3 and even W-2.1, which was inferred to be caused by the larger extent of fibre damage during the weaving process due to its less regular weave pattern than W-2.1 and W-2.3, although this has not been confirmed.

Based on the normalised results, W-3 was the highest in both tensile modulus and tensile strength. The compact weft tow placement and the angled binders resulted in less crimped warp tows which led to a higher tensile modulus. The weave architecture also had less and smaller resin rich regions which were found to be the damage initiation sites in the other weaves. Therefore the fewer damage initiation sites resulted in higher tensile strength of W-3.

Weave W-4 exhibited a large extent of non-linearity after about 0.5% strain due to the absence of straight load-carrying warp fibres. The angle-interlock binders started to straighten up during loading, which caused the non-linear behaviour and matrix failure in the resin pockets and then induced failure. Therefore W-4 had the lowest tensile properties among all of the weaves.

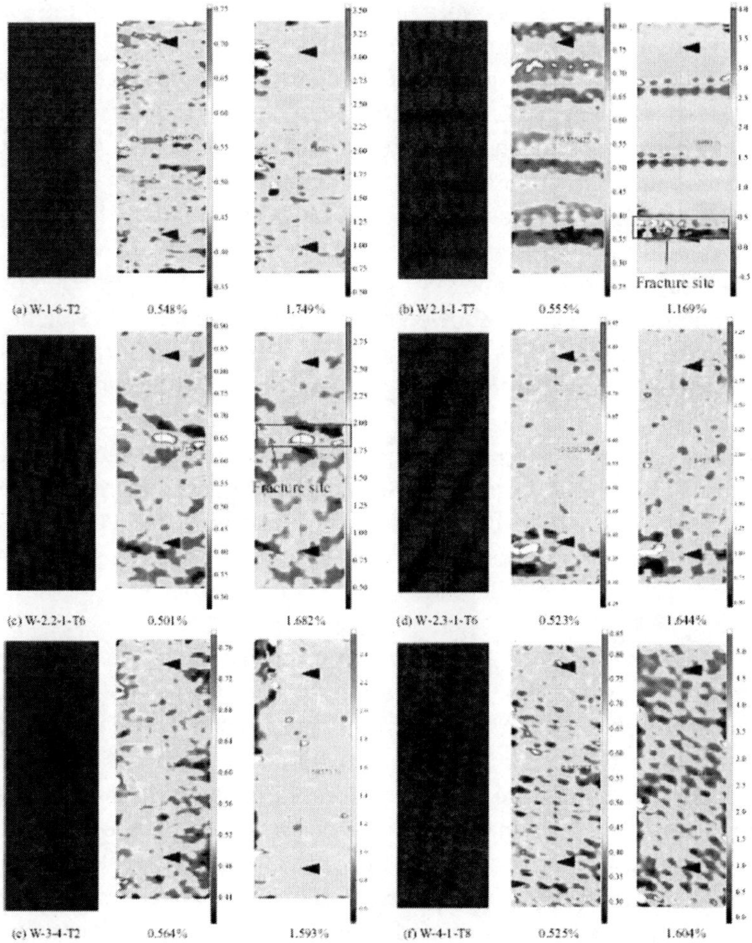

Figure 11. Axial strain (%) distributions at different strain level and the last frame before fracture. (For interpretation of the references to colour in this figure legend, the reader is referred to the web version of this article.)

Fig. 11 shows the strain distribution maps obtained at the averaged strain level of approximately 0.5% strain and at the last frame before fracture, along with the corresponding weave pattern on the left. A 50 mm long virtual strain gauge was attached on the strain map to give the averaged strain within the gauge area. As shown in the strain maps, the strain distribution was largely affected by the weave patterns. Localised high strain regions were detected near the binding points within the resin rich zones where matrix cracks were first visually observed. Two of the samples failed within the DIC monitored region and the final fracture sites are illustrated on the strain maps in these two samples (W-2.1 and W-2.2) in Fig. 11. In W-1 and W-2.1, high strain regions were detected within the resin rich channels at an average strain of around 0.5%. W-3 showed a relatively uniform strain distribution with a few strain concentration regions around the binding points. In all other weaves, the high strain regions concentrated near the binding points since there were no clear connected resin rich channels. As can be seen from Fig. 11(a) and (e), the left side of the specimens showed higher strain than the right side at lower strain level, and this difference almost vanished at higher strain levels. This was possibly caused by the misalignment between the specimen and the DIC camera. The strain caused by this misalignment was lower than the strain at higher stress level, hence became less noticeable. Since the purpose of the DIC was only to qualitatively characterise the strain distribution, this strain difference was neglected. However, more care should be taken in positioning the camera and the specimens in future DIC tests.

The binder tow straightening and matrix cracking around the binding points were visually observed on the surface of the specimens in all weave types. According to the strain distribution maps, strain concentration zones located around the binding points where the binder fibres changed directions, and therefore the matrix cracks initiated within these strain concentration regions. In all of the orthogonal weaves, the matrix cracks initiated around the binding points, propagated along the resin rich channel, and then coalesced together into a longer transverse crack across the width of the sample. In W-1 and W-2.1, warp tow debonding was visually observed through the resin channels after matrix cracking. It was also observed in W-2.1 that the matrix crack initiated inside the weft tow and propagated to the boundary of the weft tow and formed a delamination crack between the weft tow and the warp/binder tow. The stress levels at which these cracks were observed were recorded in some of the samples. The cracks appeared at the edges in W-2.1 at 489 MPa, in W-2.2 at 787 MPa, and in W-4 at 422 MPa, and surface indentation caused by the binder tow straightening was found in W-3 at 422 MPa. However, since these cracks were only detected visually, the initiation stress of these cracks were inconclusive and cannot be used for comparison.

The final tensile failure was the breakage of the warp tows, which occurred along the coalesced matrix cracks within the resin channels. Since the resin channel was curved in W-2.2 and W-2.3 as shown in Fig. 3, the final fracture surface followed the curved weft tows in these two weaves. Fig. 10 shows the cross-sections parallel to the loading direction in the vicinity of final fracture sites. It can be seen that matrix cracks within the weft tows occurred in all weave types. Clear separation between warp and weft layers was observed in W-1, W-2.1, and W-2.2.

The weaves with more compact architectures (W-1 and W-3) exhibited higher tensile properties, because the compact structures provide less crimped fibre (W-1) and fewer damage initiation sites (W-3). The weave without non-interlacing fibres (W-4) showed the lowest tensile properties since the benefit of the reinforcement lies in the fibre direction, which was not fully utilised in W-4.

4.2. Compression testing

Fig. 12 shows the typical compressive load and displacement curves of the end-tabbed samples of the six weave architectures. Similar to the tensile properties, W-4 exhibited lower compressive properties than other weaves due to the absence of non-crimp fibres in the loading direction. However, it can be seen from the load–displacement curves that the W-4 sample can still carry load at about 75% of its maximum strength while all other weaves had failed completely. It was also observed during testing that the failure process of W-4 was less catastrophic than the other weaves. The averaged compressive properties for all weaves are listed in Table 5 and the normalised results are shown in Fig. 13. As can be seen that W-3 and W-4 were the highest and lowest in compressive modulus and strength, all four orthogonal weaves had similar compressive strength, and W-2.1 had the lowest compressive properties among the orthogonal weaves. Since the compression tests were sensitive to misalignment and the length of the strain gauge (2 mm) was shorter than the length of the unit cell (up to 10 mm), the scatter of the compressive properties data was relatively high. Fig. 14 shows the fracture sections of the samples from the compressive strength tests. The main compressive failure features included matrix cracking, delamination, warp tow fracture which was the final fracture feature for all weaves except W-4.

In W-1 the delamination crack occurred in between every two layers and propagated for at least two unit cells before being impeded by the binder tows. In W-1 the binder tows were placed between every other weft tow therefore the delamination crack had to break all of the binder tows to propagate to the adjacent cells, which results in a shorter delamination crack than W-2.1.

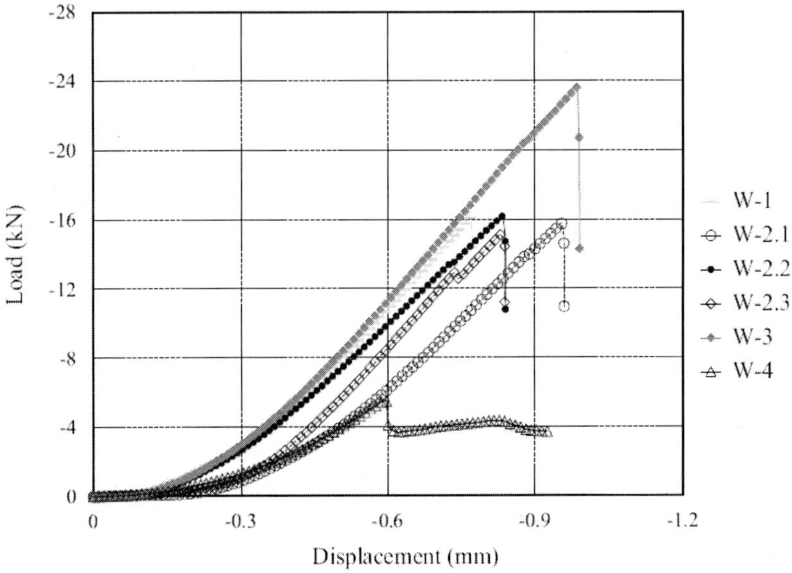

Figure 12. Typical load–displacement curves for all six types of woven composites under compression. (For interpretation of the references to colour in this figure legend, the reader is referred to the web version of this article.)

Table 5. Compressive properties.

Weave ID	Modulus (GPa)	CV (%)	Strength (MPa)	CV (%)
W-1	76.18	10.22	444.31	30.46
W-2.1	62.21	12.57	435.59	15.67
W-2.2	77.41	18.81	444.80	18.84
W-2.3	87.19	15.40	421.41	12.82
W-3	92.29	16.27	549.59	12.76
W-4	57.25	23.18	197.41	17.84

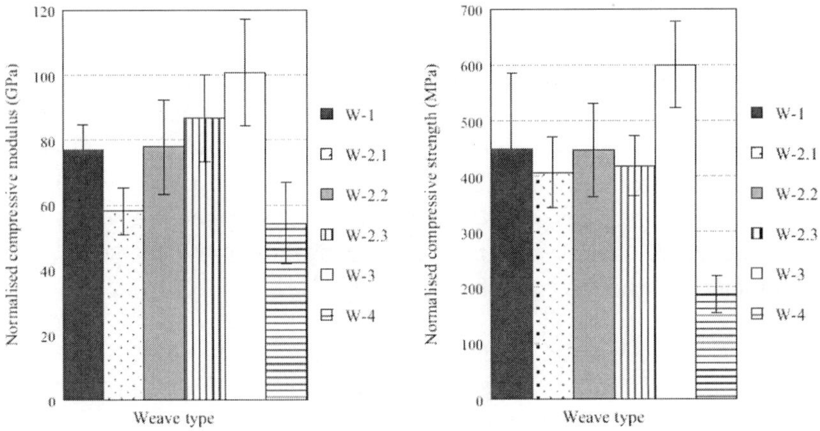

Figure 13. Normalised compressive properties.

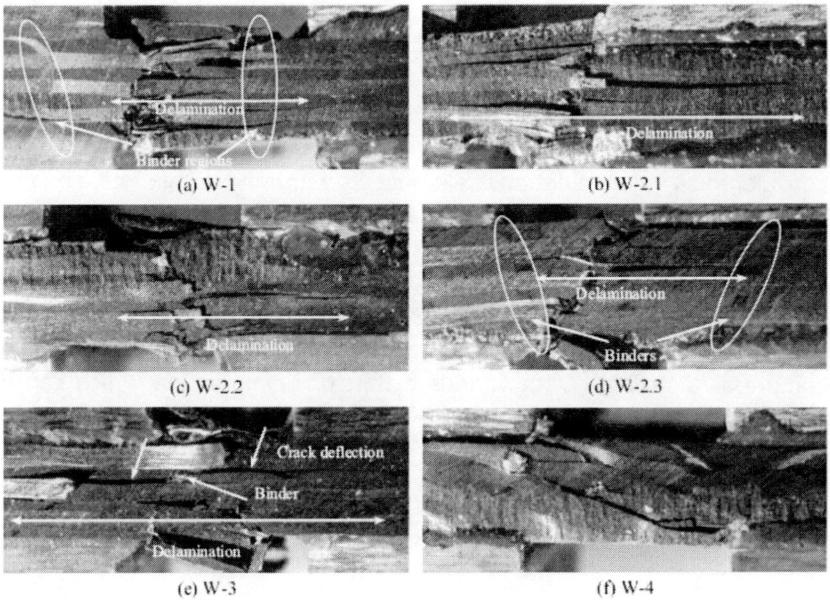

Figure 14. Failed sections of compression samples (parallel to warp direction). (For interpretation of the references to colour in this figure legend, the reader is referred to the web version of this article.)

Weave W-2.1 had longer delamination cracks since the through-thickness binder interlaced three weft tows together with no other binder inserted in between to prevent the crack growth. The delamination crack length was reduced in W-2.2 and W-2.3, which was attributed to the different binding sequence with more frequent binder insertion. Although in W-2.2 and W-2.3 the binder tow also interlaced three weft tows together similar to W-2.1, another binder tow was inserted within these three weft tows in the adjacent cell, which resisted the crack propagation in the adjacent cell and eventually reduced the crack length.

Weave W-3 also had longer delamination cracks, because the binder tows in W-3 were at an angle to the crack, which did not prevent crack propagation as effectively as the orthogonal binders. In addition, the delamination crack was deflected by the binder tow and propagated from between the first warp layer and the second weft layer to between the second weft layer and the middle warp layer, as shown by the arrows in Fig. 14.

It was visually observed in W-4 that cracks initiated at the boundary of the binder tows first, which caused separation between the interlaced tows and resulted in the loss of load-bearing capacity. Since there were no distinctive layers and all binders and weft tows were interlaced, W-4 did not have any delamination cracks. The crack shown in Fig. 14 is not considered as a delamination crack since its width did not extend along the width of the sample due to the interlacing binder in the adjacent cell.

The weave with the least crimped tows (W-3) showed higher compressive properties, however it exhibited longer delamination cracks due to the angled binder tows. The four orthogonal weaves showed similar compressive strength with various extent of delamination which can be reduced by changing the binding sequence. W-4 offered the lowest compressive properties due to the absence of non-crimp warp tows but exhibited non-catastrophic failure.

4.3. Flexure testing

The typical stress–strain curves for the six weaves under three-point bending load are plotted in Fig. 15, and the actual and normalised flexural properties are presented inTable 6 and Fig. 16. Since the span-to-thickness ratio was higher than sixteen and the displacement was larger than 10% of the span, the flexural stress was calculated based on Eq. (2) to correct the influence of excessive end-forces induced by large span-to-thickness ratio according to ASTM D790 [29]. The flexural strain was also calculated using Eq. (3).

$$\sigma_f = (3PL/2wt^2)\left[1 + 6(D/L)^2 - 4(D/L)(t/L)\right] \quad (2)$$

$$\varepsilon_f = 6Dt/L^2 \quad (3)$$

where σ_f is the flexural stress, *P* is the applied load, *L* is the support span, *w* and *t* are the width and thickness of the sample, *D* is the displacement, and ε_f is the strain at the centre of the specimen.

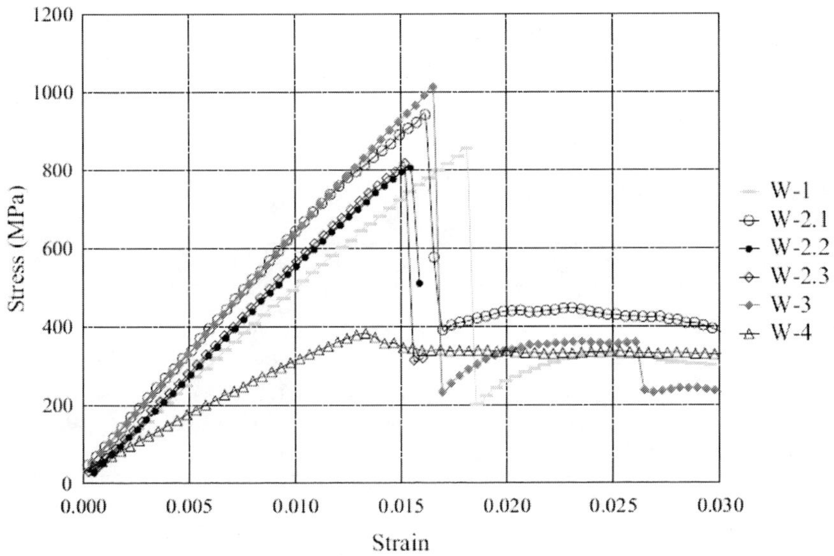

Figure 15. Typical stress–strain curves for all six types of woven composites under three-point bending. (For interpretation of the references to colour in this figure legend, the reader is referred to the web version of this article.)

In all of the weaves, the flexural modulus was lower than the tensile and compressive modulus, and the flexural strength was higher than the compressive strength even though the final flexural fracture was caused by compressive failure. Similar testing results were reported by Wang and Zhao [12] on a similar orthogonal 3D woven composite. One possible explanation for the lower flexural modulus is that the middle warp tow passed through the bending neutral plane and therefore was not fully loaded as it did under tensile loading. Therefore one of the three main load-carrying tows was not fully utilised under flexural loading, which resulted in the lower flexural modulus. The higher flexural strength might also be caused by the uneven distribution of reinforcement in the through-thickness direction. The reported flexural strength was the stress on the outer surface, which was calculated using the classical beam theory for isotropic material. Due to the inhomogeneity in the through-thickness direction, this calculated stress was not the stress in the warp tows which failed in compression. In addition, the reported compressive strength was also an averaged stress over the entire cross-section not the stress within the failed warp tows. Since there were no well-established methods to estimate the stress within a tow in a 3D woven composite, the flexural strength can not be compared directly with compressive or tensile strength.

Table 6. Flexural properties.

Weave ID	Modulus (GPa)	CV (%)	Strength (MPa)	CV (%)
W-1	48.49	8.42	886.48	7.36
W-2.1	62.22	8.87	960.12	9.54
W-2.2	55.04	4.14	805.26	11.65
W-2.3	58.57	7.70	826.52	5.74
W-3	63.32	4.04	1036.62	10.78
W-4	30.37	2.41	379.89	2.60

As can be seen from Fig. 15, all weave types exhibited a linear stress–strain behaviour at the beginning of loading, and showed some non-linear behaviour after about 0.01 strain. All of the weaves except W-4 exhibited a load drop once they reached flexure strength, which was caused by the brittle fracture of the load-carrying warp tows on the compression side of the specimen. Since W-4 did not have any straight warp fibres, therefore no catastrophic load-carrying fibre breakage occurred. However the flexural properties of W-4 were the lowest among all weaves. W-2.1 and W-3 had higher flexural properties as listed in Table 6, and were still the higher weaves in the normalised results as shown in Fig. 16 due to the angled binder. The flexural properties of W-2.1 were higher than the other orthogonal weaves while its tensile properties were relatively low. A possible explanation is that the angle of the binder tow played a more important role in bending than the warp tow waviness. The binder tow in W-2.1 was at an angle similar to W-3 rather than being orthogonal as shown in Fig. 2. Under flexural loading, the binder tow did not straighten and induce cracks as in the tensile tests, instead it acted as a "truss" and improved the bending properties. Therefore the weaves with angled binder tows (W-3 and W-2.1) showed higher flexural modulus.

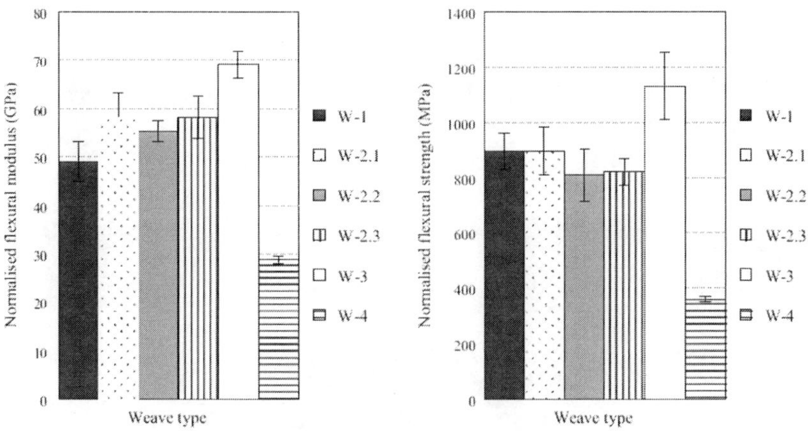

Figure 16. Normalised flexural properties.

Fig. 17 shows the failed sections of all six weave types. The failure consisted of the matrix cracking near the binding points, delamination, and warp tow fracture which occurred on the compression side under the loading point. Delamination occurred in between the first warp layer and the second weft layer in all of weaves except W-4 and the length of the delamination cracks varied between designs. W-2.1 and W-3 had the longest delamination cracks among all and relatively shorter delamination was found in W-2.2 and W-2.3.

Weave W-1 exhibited a longer delamination crack than W-2.2 and W-2.3. W-1 was designed to have the same amount of fibres in the through-thickness direction within a same area as the other orthogonal weaves. Therefore the delamination crack should break the same amount of through-thickness fibres within the same region for all orthogonal weaves. In addition, the through-thickness binder tows occurred in between every two weft tows in W-1, which was more frequent than in other orthogonal weaves. However, W-1 used 1k yarns as binders while all other orthogonal weaves used 3k yarns which required more energy to break. It is shown in Fig. 17 that the binder tows in W-1 were fractured by the delamination crack while in W-2.2 and W-2.3 the binder tows deflected the crack and were not fractured by delamination. In addition, the through-thickness fibres were placed closely and evenly along the weft direction. Therefore the delamination crack would have an even propagation front, which promoted a steady crack growth and hence a longer crack.

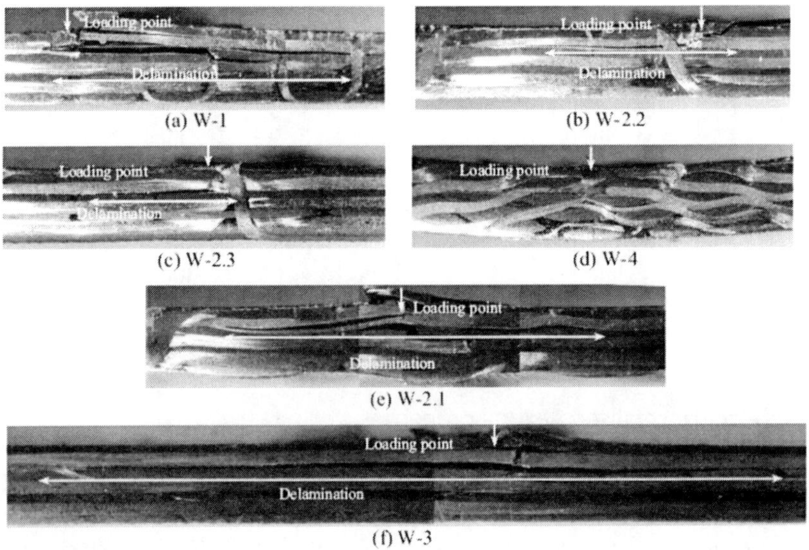

Figure 17. Failed sections of flexure samples (parallel to warp direction). (For interpretation of the references to colour in this figure legend, the reader is referred to the web version of this article.)

For W-2.1, although it had the same number of binder tows as W-2.2 and W-2.3, it did not have a wide distribution of the through-thickness binder fibres. The through-thickness portions of the binder tows in W-2.1 were all placed within the same binding lines along the same weft tows. Therefore in between two adjacent binding points along the binder, there were no through-thickness fibres resisting delamination, which resulted in a longer delamination. While in W-2.2 and W-2.3, the delamination crack was arrested or retarded by the binder tows in the adjacent cells.

In W-3, the binder tows were at an angle, so they did not prevent crack propagation as effectively as the orthogonal binders due to their shorter and angled through-thickness portions. Also the spacing between two binding points in the warp direction was longer in this angle-interlock weave than in the orthogonal weaves. This larger spacing also caused a longer delamination crack since the delamination can propagate further before getting impeded by the binder tows at the next binding point.

Weave W-4 had the lowest flexural properties among all weave types due to the absence of non-interlacing warp tows. However, W-4 exhibited a more gradual failure process while all other weaves failed catastrophically.

The weave with angled binder tows (W-3 and W-2.1) showed higher flexural properties, however they experienced longer delamination cracks. The angled binder was inferred to increase the flexural properties by acting as a truss, however further tests are required to confirm this. The weave without non-interlacing fibres (W-4) showed the lowest flexural properties but failed progressively.

5. CONCLUSIONS

The main objective of this study was to understand the effect of weave architecture and binder placement on the behaviour of the 3D woven composites. Six types of 3D woven carbon fibre composites with the same types of fibres were manufactured using a traditional narrow fabrics weaving loom and resin transfer moulding technique. The tensile, compressive, and flexural modulus and strength in the warp direction of all six weave architectures were experimentally characterised in this work, and the influence of weave architectures on the behaviour of 3D woven composites was revealed. Based on the experimental observations and analysis the following conclusions are made:

1) The waviness of the tows were determined by the weave architectures during the weaving process even though all warp and weft tows were ideally designed to be non-crimp. A compact binding sequence resulted in less crimp in the warp tows such as in W-1 and W-2.2. Longer weft

flows with the two adjacent binder tows having opposite binding movement caused the in-plane waviness of the weft tows.

2) The tensile properties of the 3D woven composites were affected by the waviness of the load-bearing fibres. One-by-one orthogonal W-1 and angle-interlock W-3 had higher tensile modulus (78.26 GPa and 79.81 GPa) and tensile strength (1370.92 MPa and 1276.24 MPa) due to their less crimped warp tows. The angle-interlock W-4 showed non-linear stress–strain behaviour due to the absence of non-interlacing warp fibres and the straightening effect of the binders. Two-dimensional strain maps obtained from a DIC system showed strain concentration zones around the binding points where matrix cracks were visually observed.

3) Weave W-3 had the highest compressive properties but also longer delamination crack. All of the orthogonal weaves had similar compressive strength, however W-2.1 had lower compressive modulus due to its crimped warp tows. Delamination occurred to all samples with straight warp tows, and was longer in W-3 and W-2.1 which had longer distance between two binding points. The delamination length can be reduced by changing the binding sequence to one with shorter binder spacing.

4) The flexural properties were found to be higher in W-2.1 and W-3, which was inferred to be caused by the angled binder acting as a truss. However the delamination cracks were longer in these two weaves due to binding arrangement. More tests are required to further understand the flexural strengthening mechanism.

5) Overall, the angle-interlock W-3 exhibited the best performance under all three loading conditions however it had the longest delamination crack among all weaves under both compression and bending loading.

ACKNOWLEDGEMENTS

This research project is supported by the Aeronautical and Automotive Engineering Department of Loughborough University and M.Wright & Sons Ltd. The help and support of the department staff and M.Wright & Sons Ltd are gratefully acknowledged.

REFERENCES

1. Tong L, Mouritz AP, Bannister MK. 3D fibre reinforced polymer composites. Oxford: Elsevier Science; 2002. p. 107–36. .

2. Mouritz AP, Cox BN. A mechanistic interpretation of the comparative in-plane mechanical properties of 3D woven, stitched and pinned composites. Compos Part A: Appl Sci Manuf 2010;41(6):709–28.

3. Bogdanovich AE, Karahan M, Lomov SV, Verpoest I. Quasi-static tensile behavior and damage of carbon/epoxy composite reinforced with 3D noncrimp orthogonal woven fabric. Mech Mater 2013;62:14–31.

4. Baucom JN, Zikry MA. Low-velocity impact damage progression in woven eglass composite systems. Compos Part A: Appl Sci Manuf 2005;36(5):658–64.

5. Chen F, Hodgkinson JM. Impact behaviour of composites with different fibre architecture. Proc Inst Mech Eng Part G: J Aerospace Eng 2009;223(7): 1009–17.

6. Bibo GA, Hogg PJ. The role of reinforcement architecture on impact damage mechanisms and post-impact compression behaviour. J Mater Sci 1996;31(5):1115–37.

7. Brandt J, Drechsler K, Arendts FJ. Mechanical performance of composites based on various three-dimensional woven-fibre preforms. Compos Sci Technol 1996;56(3):381–6.

8. Ivanov DS, Lomov SV, Bogdanovich AE, Karahan M, Verpoest I. A comparative study of tensile properties of non-crimp 3D orthogonal weave and multi-layer plain weave e-glass composites. Part 2: comprehensive experimental results. Compos Part A: Appl Sci Manuf 2009;40(8):1144–57.

9. Chou S, Chen H-C, Chen H-E. Effect of weave structure on mechanical fracture behavior of three-dimensional carbon fiber fabric reinforced epoxy resin composites. Compos Sci Technol 1992;45(1):23–35.

10. Stig F, Hallstrom S. Assessment of the mechanical properties of a new 3D woven fibre composite material. Compos Sci Technol 2009;69(11-12): 1686–92.

11. Cox BN, Dadkhah MS, Morris WL, Flintoff JG. Failure mechanisms of 3D woven composites in tension, compression, and bending. Acta Metall Mater 1994;42(12):3967–84.

12. Wang Y, Zhao D. Effect of fabric structures on the mechanical properties of 3D textile composites. J Ind Text 2006;35:239–56.

13. Dai S, Cunningham PR, Marshall S, Silva C. Mechanical behaviour of 3D woven composites under tension, compression and bending. In: Proceedings of ICCM- 19, CACSMA, Montreal, Canada, 2013, p. 2812–2820.

14. Cox BN, Dadkhah MS, Morris WL. On the tensile failure of 3D woven composites. Compos Part A: Appl Sci Manuf 1996;27(6):447–58.

15. Gu H, Zhili Z. Tensile behavior of 3D woven composites by using different fabric structures. Mater Des 2002;23(7):671–4.

16. Lomov SV, Bogdanovich AE, Ivanov DS, Mungalov D, Karahan M, Verpoest I. A comparative study of tensile properties of non-crimp 3D orthogonal weave and multi-layer plain weave e-glass composites. Part 1: Materials, methods and principal results. Compos Part A: Appl Sci Manuf 2009;40(8):1134–43.

17. Callus PJ, Mouritz AP, Bannister MK, Leong KH. Tensile properties and failure mechanisms of 3D woven grp composites. Compos Part A: Appl Sci Manuf 1999;30(11):1277–87.

18. Guess TR, Reedy Jr ED. Comparison of interlocked fabric and laminated fabric kevlar 49/epoxy composites. J Compos Technol Res 1985; 7(4):136–42.

19. Cox BN, Dadkhah MS, Inman RV, Morris WL, Zupon J. Mechanisms of compressive failure in 3D composites. Acta Metall Mater 1992;40(12): 3285–98.

20. Kuo WS, Ko TH. Compressive damage in 3-axis orthogonal fabric composites. Compos Part A: Appl Sci Manuf 2000;31(10):1091–105.

21. Kuo WS, Ko TH, Chen CP. Effect of weaving processes on compressive behavior of 3D woven composites. Compos Part A: Appl Sci Manuf 2007;38(2): 555–65.

22. Jin L, Niu Z, Jin BC, Sun B, Gu B. Comparisons of static bending and fatigue damage between 3D angle-interlock and 3D orthogonal woven composites. J Reinf Plast Compos 2012;31(14):935–45.

23. Kuo WS. The role of loops in 3D fabric composites. Compos Sci Technol 2000;60(9):1835–49.

24. Adanur S, Tam CA. On-machine interlocking of 3D laminate structures for composites. Compos Part B: Eng 1997;28(5-6):497–506.

25. University of Nottingham. TexGen, on-line (August 2013).

26. ASTM International. ASTM D3039/3039M standard test method for tensile properties of polymer matrix composite materials; 2008.

27. Adams DO. A comparison of CEN and ASTM test methods for composite materials; June 2004.

28. ASTM International. ASTM D695 standard test method for compressive properties of rigid plastics; 2010.

29. ASTM International. ASTM D790 standard test methods for flexural properties of unreinforced and reinforced plastics and electrical insulating materials; 2010.

30. Zhou G, Davies GAO. Characterization of thick glass woven roving/polyester laminates: 1. Tension, compression and shear. Composites 1995;26(8): 579–86.

Chapter 4

Mechanical Analysis of Woven Fabrics:The State of the Art

Savvas Vassiliadis[1], Argyro Kallivretaki[1], Dimitra Domvoglou[1] and Christofer Provatidis[2]

[1] Department of Electronics, Technological Education Institute of Piraeus,, Greece

[2] School of Mechanical Engineerin, National Technical University of Athens, Greece

1. INTRODUCTION

The automation and integration of processes in the textile industry is dictated by the increasing need to offer specialized products at optimum quality and low cost, satisfying at the same time the fast cycles of fashion trends or in the case of technical applications the delivery of products of high qualiy and of exact properties. Under these premises, computer engineering tools, such as computer-aided engineering (CAE) and computer-aided design (CAD), have recently gained attention. The revolutionary role of CAE and CAD tools in the textile industry is the guaranty that the final product meets the set specifications, optimizing thus the quality control procedure. Moreover, the prediction of the properties and the aesthetic features of the product before the actual fabrication can essentially benefit the textile research community [Hu and Teng, 1996]. Especially nowadays that textile materials can be used for the production of a wide range of technical products, such as reinforcements in composites for aerospace or marine applications or textiles for medical applications, the prediction of the end-product's mechanical properties is of major importance. Furthermore, the textile raw materials are processed under low-stress conditions and it is thus reasonable to assume that the knowledge of the possible modifications introduced via the manufacturing process is necessary for the final product realization (Hu, 2004).

Textiles are flexible, anisotropic, inhomogeneous, porous materials with distinct visco-elastic properties. These unique characteristics makes textile structures to behave essentially different compared with other engineering materials. Moreover, textiles are characterized by an increased structural complexity. Their properties mainly depend on a complicated combination of their structural units and their interactions. The complicated nature of the textiles' mechanics makes them ideal candidates for a mechanical analysis using computer-based methods.

This paper focuses on the investigation of the modeling attempts of woven fabrics. The woven fabrics' weave patterns as well as the deformation mechanisms of their consistent yarns make these structures modelling extremely challenging (Parsons et al., 2010). An extended literature review of the computational models for the deformation of woven fabrics is presented. Based on these models, the difficulties towards a comprehensive model for textile structures are highlighted. Taking into account the existent literature, the perspective of developing a widely accepted integrated CAE environment for textiles (Hearle, 2006), is also extensively discussed.

2. TEXTILE STRUCTURES AND THEIR MECHANICAL BEHAVIOR

Since this study focuses on the investigation of the existent woven fabrics' simulation techniques, some introductory remarks concerning the basic structural units of these unique substrates, are thought to be extremely useful. Textile fabrics are made of interlaced yarns which consist of the basic element of every textile product, the fibres. Fabrics are classified according to their manufacture process as knitted, woven and non-woven.

The computational representation of textiles is hindered by the geometrical complexity of the textiles' basic structural units (fibres and yarns) as well as of the fabrics' weaving and knitting patents. The aforementioned characteristics of the textiles woven fabrics result in complicated deformations even in cases of simple loading. For example, the tensile deformation of a spun yarn corresponds to the superposition of bending, tensile and compression of the constituent helically arranged fibres. Furthermore, contact phenomena, as sticking and sliding interaction, should be also taken into account in the mechanical deformation analysis increasing thus further the complexity of the mechanical study.

Fabric mechanics study often leads to the introduction of models with simplifying assumptions. The yarn, which is usually assumed as a homogeneous material, is considered as the basic structural unit of the fabrics. The elastic properties of the homogeneous yarn result from the elastic properties of the fibres and include the non-linear structural synergy of them within the yarn body. Even if the yarns are assumed as homogeneous materials, the contact phenomena dominate the deformation procedure of the fabrics. Actually, the friction effects support the stability of the textile structures. The contact phenomena have also a great significance for the stress and strain distribution in a fabric subjected to deformation. The friction energy losses appear during the load transferring along threads. Thus, very often, uneven load distribution appears within the textile structures.

Due to the large deflection effects and the nonlinearity of the textile structures' deformation phenomena, the fabrics mechanics study requires special attention. The relative large deformation of the fabrics arises from the flexibility of the textile fibres and yarns as well as from the structural details and the way of the load application. The yarns present high deformability which results from the low values of packing factor (the ratio of the fibres volume to the total volume of the yarn). The air trapped between the fibres is easily removed during the axial loading imposing the reduction of the apparent yarn cross sections and thus the high deformation of the yarns which is obviously transferred to the fabrics. Moreover the pattern of the fabrics itself and especially the structure of the fabrics, supports the development of high deformations. From the structural point of view the fabric pattern can be considered as a multi-body system of yarns. The tensile deformation of the fabric corresponds to the synthesis of two processes, the bent yarns' straightening and their subsequent elongation. The first process dominates in the lower loading stage and the second process appears upon the increase of the load. Thus the load-deflection curves of a textile structure subjected to tensile deformation is strongly nonlinear. The nonlinearity is also supported from the change of the contact status between the yarns, the large deflection effects observed even within the unit cell of the fabric and finally the material nonlinearities

2.1. Technical Applications Of Textiles

Although conventional textiles are primarily used for clothing, the use of a variety of raw materials as well as the development of new manufacturing processes led to a considerable expansion of their possible applications. The importance of aesthetic and decorative characteristics of textiles has been decreased by the new materials' performance and functionality. The growing recognition of the textiles potentials led to revolutionary new technical applications which according to Techtextiles (the international Trade exhibition for technical textiles) are (Horrocks, 2000):

- Agrotech: agricultural (nonwoven for wind protection)
- Buildtech: building and construction (awning, concrete reinforcements)
- Clothtech: clothing (garments)
- Hometech: household (curtains, wall covering)
- Indutech: industrial applications textiles (filters)
- Medtech: medical (bandages, sutures)
- Mobitech: mobility (ropes, seat covers)
- Oekotech: eco-friendly textiles (recyclable composites)
- Packteck: packaging (nets, wrappings)
- Protech: protection (bullet-proof jackets, uniforms)
- Sportech: sports and leisure (carbon-fibre composites for racquet frames)
- Geotech: geotextiles (nonwovens for drainage, reinforcement)

Over the last decades there is also an intensive need for high-tech materials with "life functions". Consequently, research interest has been moved towards the development of textile-based structures which change their properties in response to an external stimulus, offering products with increased functionalities. The so called "intelligent", "smart" textiles in conjunction with the wearable electronics usually consisting of electronic modules incorporated into textiles, support activities in military, telemedicine or rehabilitation (Rossi et al. 2006; Tang, 2007; Cho et al., 2009). Textiles' flexibility, indicative of the wearer's comfort, makes them ideal candidates for interfaces in contact with the human skin. Based on these assumptions, a large number of wearable electronic systems have been developed (Dunne et al., 2005; Xu et al., 2008; Tognetti et al., 2006). Smart textiles development requires the synergistic action of different disciplines such as textile science and engineering, natural sciences, material science, mechanical engineering, electrical and computer engineering and informatics, making this promising research area extremely challenging. Furthermore, the attention attracted by this dynamic sector of textile research is thought to make a contribution towards a cost effective commercialization of innovative textile-based products aiming in the improvement of people's quality of life.

(a) (b) (c)

Figure 1. Technical applications of textiles (a) Fabric for solar protection used in the Paul Klee museum, (**soltis-textiles, 2011**), (b) DuPont's Kevlar® XP™ developed for hard armor applications, (**DuPont, 2011**), (c) Space suit developed by ILD Dover (**ilddover, 2011**)

(a) (b) (c)

Figure 2. Smart textiles (a) Interactive Led dance shoe by Moritz Waldemeyer (**waldemeyer, 2011**) (b) Luminate textiles by Phillips (**smarteconomy, 2006**), (c) Led Dress by Cute Circuit (**crunchwear, 2010**)

3. MECHANICAL MODELLING OF THE TEXTILE STRUCTURES

3.1. Classification Of The Modelling Approaches

During the last decades, several methods were adopted for the mechanical modelling and analysis of the textile structures. A basic classification, according to the modelling method used, divides them into the analytical and numerical or computational approaches. The dominant engineering design culture played important role for the development and the succession of these approaches. Classical modelling methods find in textiles an attractive application field. Another essential classification of the modelling of the textile structures is made according to the scale of the model. There is micromechanical, mesomechanical and the macromechanical modelling. The micromechanical modelling stage focuses on the study of the yarns, tows even fabrics taking into account the structure, orientation and mechanical properties of the constituent fibres. The mesomechanical modelling, on the other side, studies the mechanical characteristics of the fabric unit cell considering the yarns as homogenous structures. Finally the macromechanical modelling stage is reffered to the prediction of mechanical performance of the fabric in complex deformations, as drape, studing the fabric as a continuum material.

Although the mentioned modelling stages were developed as distinct analysis approaches, their integration in a compound modelling approach was directly rised. Thus the textile society implemented a modelling hierarchy (Takano et al., 1999; Lomov et al. 2004; Bogdanovich, 2006) based on three modelling scales: the micromechanical modelling of yarns, the mesomechanical modelling of the fabric unit cell and the macromechanical modelling of the fabric sheet (Figure 3).

According to the integrated textile modelling concept, the only inputs in the total design procedure are the fibre properties, the yarn structure and the fabric structure. In the first modelling stage, the fibre properties and the yarn structure (yarn type, number of fibres, orientation) are introduced as input parameters for the mechanical analysis of the yarn and the calculation of the yarn properties. Then the yarn properties are transferred in the second modelling stage. The selection of the required yarn properties and their attribution in the modelled yarns corresponds to a homogenization procedure that connects the two individual stages. Moreover the woven fabric structure is introduced in the mesomechanical modelling stage. At the current stage the yarns are represented as continuum structures and the analysis is limited on the study of the fabric unit cell. Then a second homogenization stage is required for the connection of the second and the third modelling stage, defining the required properties of the unit cell and their attribution in the continuum fabric models. Finally the macromechanical modelling stage based on the generation of simplified structure (usually continuum material) predicts the mechanical performance of of extended fabric pieces in complex deformations. Each individual modelling procedure such as their interface presents significant obstacles.

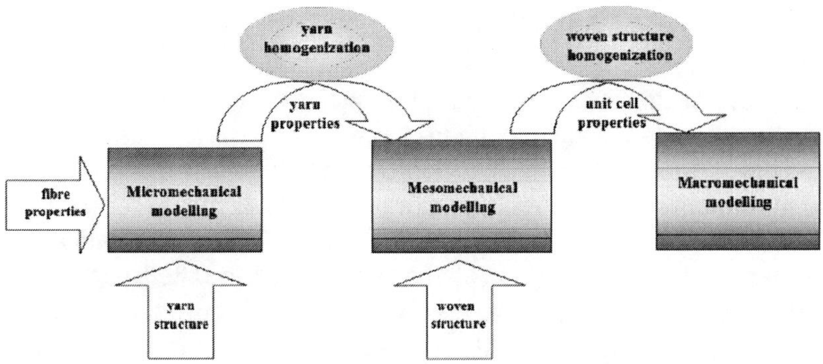

Figure 3. Integrated textile modelling.

3.2. Classification Of The Deformations

The substantial difficulty arising in textile mechanics consists in the calculation of the superposed deformations in the microscopic scale. Even a simple deformation of the woven fabric, for example a tensile deformation, incures a complex deformation mechanism of straightening, tensile, compression and sliding of yarns in the mesoscopic scale and respective deformations of fibres in the microscopic scale. Moreover the percentage of the deformation increases when it is referred to the microscopic scale since the subjected structures are smaller. For example, a 5 % shear deformation of a fabric could impose a huge displacement of the constituent fibres. Thus a simple deformation in the macroscopic scale corresponds to complex deformations in the microscopic scale. However the classification of the deformations is based on the macroscopic level. Thus the tensile, shear, bending and compression of the fabric sheet are considered simple deformations. The complex deformation of fabrics is mainly referred to the drape test. The performance of a fabric in drape is very interesting for the aesthetic effects and the dynamic functionality. The fabrics have the ability to undergo large, recoverable draping deformations by bending in single and double curvature providing a sense of fullness and a graceful appearance. Especially when the fabrics are used as reinforcement materials for the construction of composites, drape is very important since it determines the formability of the fabric in the matrix. The drapeability of the fabric reinforcement offers the advantage of bending around double-curvature mould producing complex shaped composite parts.

4. ANALYTICAL MODELING

The first mechanical modelling and analysis attempts of the textile structures started about a hundred years ago. The earliest publication probably is from R. Haas in the report of the National Advisor Committee for Aeronautics in 1918 (Haas, 1918). It is worth to mention that NACA is the early form of the today's NASA of the US. This publication is the translation from the German of the original article dated back on 1913 appeared in a German Journal. The work of Haas is of great importance. Although it is the first known, it is characterized by its integrated character. It brings together the theoretical and practical aspects up to the testing and application topics. However the work of Haas remained unknown for a long perod while the work of Peirce (Peirce, 1937) was considered as the starting reference for the analytical mechanical modelling of the textile structures. The researchers focused on the application of the existent analytical methods already used in other sectors of engineering. The main characteristic is the balance between the simplifications introduced and the precision of the modelling. The energy methods and the elastica theory are dominating in these attempts.

4.1. Micromechanical Modelling Of Simple Deformations

In the field of the analytical modelling of the textile yarns, several investigations focused on multi-filament twisted yarns. Purpose of these investigations was the prediction of the response of a twisted yarn when subjected to a certain deformation. It was supposed that the mechanical parameters such as the load-elongation curve of the constituent fibres, the twist density, the initial specific volume etc are given. The analysis focused on the correlation of the macroscopic distortion of the yarn with the microscopic response of the constituent fibres. A basic challenge in the modelling of the yarns is the balance between the realistic formulation and the idealization required for a theoretically treatable model. In the most cases the yarn was considered as being made of continuous filaments of circular cross-sections and constant linear density along their length. All the fibres were assumed to have identical properties and to be perfectly elastic. The cylindrical-helix model of Hearle et al. (Hearle et al., 1959), the conical-helix model of Önder and Başer (Önder & Bacer, 1996) and the statistical model of Komori (Komori, 2001) approached the yarn mechanical modelling from different points of view, depending on the considered alignment of the fibres. The tensile, bending and torsional behaviour of the yarns were approached using the force, the stress-analysis and the energy methods (Backer, 1952; Platt et al., 1959; Freeston & Schoppee, 1975; Choi & Tandon, 2006; Park & Oh, 2006).

4.2. Mesomechanical Modelling Of Simple Deformations

Starting point for the analytical modelling of woven fabrics was the uniaxial/biaxial deformation of the plain woven structure. The proposed approaches were based on three principal underlying geometrical models of plain weave (Dastoor et al., 1994). The “flexible thread” model of Peirce (Peirce, 1937) assumed the yarns infinitely flexible, incompressible and inextensible, without bending rigidity and having circular cross-sections (Figure 4, Figure 5). The analytical transcendental equations proposed by Peirce for the systematic description of his model cannot easily give a solution. Thus graphical and nomographic tools were presented in order to support the users. Peirce’s model has been modified later towards a better representation of the real fabric structure. Thus the assumptions of the race-track (Figure 6) or elliptical (Figure 7) yarn cross-sections (Kemp, 1958; Olofsson, 1964b) were adopted for the fabric modelling. The concept of the elastica model (Peirce, 1937), in continue, introduced the yarn bending rigidity in the analysis. According to this model the shape of yarn axis can be obtained by treating the yarns as elastic slender rods subjected to transverse point forces, equidistant but alternating in direction. In general, the mentioned models and their later modifications used the equilibrium, energy or elastica method for the mechanical analysis.

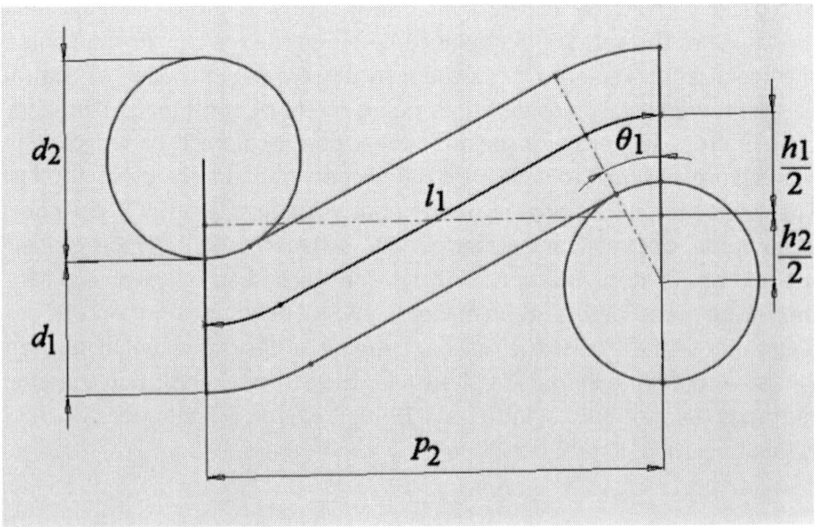

Figure 4. Plain woven geometry proposed by Peirce.

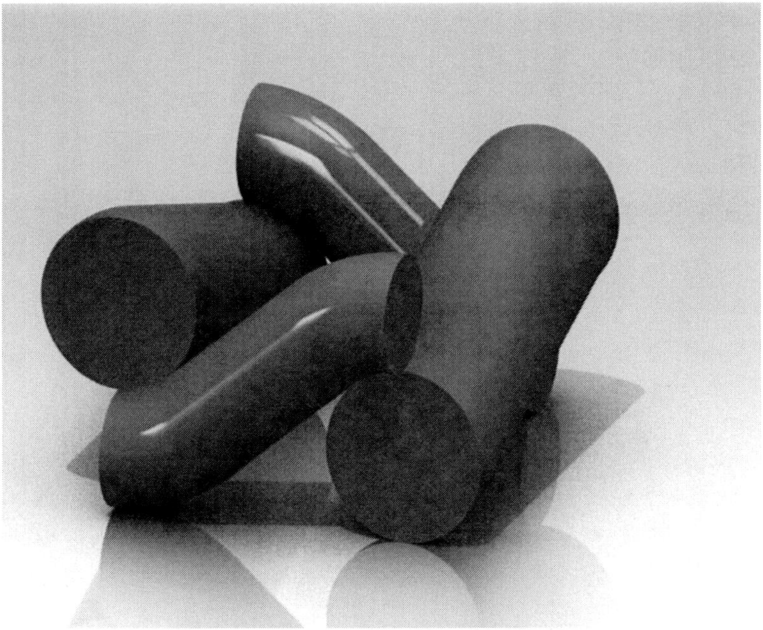

Figure 5. Representation of woven model proposed by Peirce.

Figure 6. Representation of woven model proposed by Kemp.

Figure 7. Representation of elliptic model proposed by Olofsson.

An approach including the effect of crimp and yarn extension, based on a flexible thread model was proposed by Freeston et al. (Freeston et al., 1967). The yarns were assumed as homogenous, linear elastic materials with linear work-hardening. An elastica model declining from the assumption of the standard shape yarn cross-section was published by Olofsson (Olofsson, 1964a). The shape of the cross-section of the yarns was considered as a function of the forces acting on them and the degree of set. The mathematical analysis was given on equilibrium conditions, on stress-strain relationships in extension and compression and on energy in bending. The effect of fabric set was included, also, in the work of Grosberg and Kedia (Grosberg & Kedia, 1966; Grosberg, 1966). They adopted an energy method on small deformation for the investigation of the initial load extension modulus of completely relaxed woven fabrics, while the yarns were assumed inextensible and incompressible. Another approach based on the elastica theory including linear extensibility of the yarns was given by Dastoor et al. (Dastoor et al., 1994). They assumed the yarns to be homogeneous, weightless slender rods, frictionless and undeformed by shear forces. In addition the yarns were considered as having circular section which does not deform under external forces. A computational implementation was adopted for the solution of the equilibrium equations.

The large biaxial deformation of partially and completely set plain woven fabrics was presented by Huang (Huang, 1979b; Huang, 1979a). His approach was based on the elastica model of yarns in the undeformed fabric and the combined action of extension and bending was considered for the fabric deformation. The introduction of bilinear moment-curvature relation (due to the sliding of the fibres within the yarn) in combination to the contact deformation of the yarns increases the reliability of the study. The "sawtooth" geometrical model was proposed by Kawabata *et al.* (Kawabata et al., 1973). The mechanical analysis was based on the force equilibrium and the displacement of the warp and weft yarns in the thickness direction of the fabrics at the contact point of the crossing threads. Although the geometrical representation of the unit cell was approximant, the deformation effect at the cross-over points was taken into account. Most of the models described assume an unrealistic invariable cross-sectional yarn shape along the yarn path, where Gong et al. (Gong et al. 2010), in a recent study moves towards a more realistic representation of woven yarns, suggesting an ellipse model with a variable yarn cross-sectional shape based on the various parameters, including fibre type, yarn count, yarn twist factor and cover factor. An alternative geometric model of woven fabric, based on the yarns' packing density as well as general fabric data, has been suggested by Dolatabadi and Kovař (Dolatabadi & Kovař, 2009).

4.3. Mesomechanical Modelling Of Complex Deformations

The concept of the complex deformations on a mesomechanical scale is extremely marginal. It is almost impossible to simulate on the scale of the unit cell the effects occurring during the drape of a fabric. The so called mesomechanical models for the complex deformation of the fabrics mainly refer to the bending behaviour of the fabrics. The first study in complex deformations of fabrics was conducted by Peirce (Peirce, 1937). He proposed an energy method for the analysis of 2D fabric bending. The analysis was based on the calculation of the change of the strain energy of the unit cell after the bending deformation. For the analysis the yarns were assumed to be of circular cross-section and incompressible and distributed forces were considered at the cross-sections of the yarns. Many researchers (Behre, 1961; Dahlberg, 1961; Lindberg et al., 1961; Abbott et al., 1971; Abbott et al., 1973) studied and reported the nonlinear nature of bending and shear properties. The approach adopted by Grosberg

(Grosberg, 1966) incorporated the effects of friction into the strip 2D bending analysis. Many relative research actions were carried out in continue contributing to the understanding of drape to some extent. But the 2D drape assessment cannot fully reflect the more complex 3D double curvature deformations of drape (Lo et al., 2002).

Shanahan et al. (Shanahan et al., 1978) accented the necessity of the complete drape treatment based on the structural mechanics shell theory. They also defended the consideration of anisotropic constitutive laws for the fabric sheet. Amirbayat and Hearle (Amirbayat & Hearle, 1989) used aspects of the shell theory in their theoretical investigation of the complex buckling. They correlated the drape shape with the bending, membrane and potential energies. From their investigation they concluded that drape is also influenced by other parameters such us the full set of anisotropic in-plane membrane, out-of-plane bending, cross term elastic constants, and the nonlinearity of the materials behaviour.

4.4. Macromechanical Modelling Of Complex Deformations

Many publications appeared in the past dealing with the macromechanical modelling of the complex deformations of the fabrics. For many years this specific area has concentrated the interest of many very important researchers. The most representative of them are referenced below.

An approach of the elastica theory for the analysis of complex deformations of fibres and fibre assemblies has been proposed by Konopasek (Konopasek, 1980a, 1980b, 1980c). It was based on the concept of planar and spatial elastica as developed respectively by Euler and Kirchhoff. Phenomena corresponding to the nonlinear behaviour of material, friction-elasticity, elastic-plasticity, and visco-elasticity were introduced in the analysis. The planar elastica theory was applied for the analysis of the large deflections of a yarn in a plane and the cylindrical bending of a fabric treated as sheet material. The spatial elastica was applied in the analysis of fibre buckling and crimp. The solution of the system of the resulted nonlinear differential equations was supported by computational tools.

An alternative approach to the theoretical mechanics of static drape of fabrics based on the differential geometry of surfaces was published by Lloyd et al. (Lloyd et al., 1996). They developed a computationally convenient implementation of the theoretical mechanics of fabrics. The fabrics themselves were treated as 2D continua represented by a surface without considerable thickness embedded in the 3D Euclidean space. The mechanical properties of the fabric were assigned to the model. The shape of the surface was described for both the deformed and the undeformed state by the means of the differential geometry of the surface. The strain values were deduced from the differences in the differential geometry expressions for the two extreme states. The strain values were correlated to the applied forces by the constitutive equations that express the mechanical properties of the material.

The differential geometry of surfaces for the dynamical modelling of fabric deformations was used for the approach of the problem by J. and R. Postle (Postle & Postle, 1996). The surface was considered as a series of twisted curves generated into the 3D Euclidean space. The differential geometry parameters incorporated the mechanical properties of the material (fabric) relating these mechanical properties to the changes in curvature as the surface was transformed into another surface. The deformation of the surface from the initial state to the final was mathematically modelled using the concept of homotopy. Bäcklund transformations were chosen for the solution of the nonlinear partial differential equations of the dynamic system.

Trying to combine the theoretical study to the experimental knowledge, Stump and Fraser (Stump & Fraser, 1996) analyzed the drape of a circular fabric sample over the circular disk of the drapemeter. They proposed an elastic ring-theory model of the draped fabric and used an energy analysis associated with the various large post-buckled deformations of the ring. Aim of their investigation was the study of the ability of the fabrics to present different configurations when they are draped under exactly the same conditions. The explanation of this ability was based on the calculation of the energy that corresponds to the various symmetric configurations.

4.5. Evaluation Of The Analytical Approaches

The review of the literature of the analytical methods for the mechanical analysis of textile structures demonstrates the absence of a successful globally accepted technique suitable for the textile design. The basic drawbacks of the analytical methods result from the simplifying assumptions implemented in order to generate a low-complexity geometry and mechanical problem. Thus the two-dimensional approach for the mesomechanical modelling, the attribution of isotropic elastic properties in the yarn models, the assumption of linear and isotropic properties in the macromechanical model introduces significant inaccuracies in the textile modelling. However the basic disadvantage of analytical approaches is the difficulty in handling in respect to the time consumption, the application field in terms of structures and materials, and the integration of the individual stages (micro, meso, macro). On the other side the analytical approaches accented the modelling difficulties of textile mechanics, the basic considerations and roadmap for an integrated design procedure.

5. NUMERICAL MODELING

The enormous computational power arose from the development of the computer systems and the expansion of advanced commercial software codes for the analysis of mechanical problems was guiding the textile design towards the numerical approaches. Mainly the Finite Element (FEM) and Boundary Element Method (BEM) were used for the mechanical modelling of the textile structures (Hu & Teng, 1996).

5.1. Micro- And Mesomechanical Modelling Of Simple Deformations

The first attempts in the computer based mesomechanics of textiles dealt with the 2D and 3D representation of the plain woven structure. The geometry proposed by Haas and then by Peirce was the starting point for the solid geometrical modelling since the numerical techniques succeed the solution of the complex system of equations. Keefe et al. (Keefe et al., 1992) based on Peirce's geometry presented the solid model of the plain woven fabric. They also extended the model for various compactions and fabric angles. Later comparative studies examined the accuracy of the geometrical models for use in the numerical modelling of fabrics (Provatidis & Vassiliadis, 2002, 2004,Provatidis et al., 2005).

The first studies in mechanical analysis of textiles focused on the tensile deformation of the plain woven unit cells. The initial use of computational methods in textile mechanics was oriented towards the numerical solution of the complex analytical expressions. The use of numerical methods, as FEM, BEM etc, for the achievement of a rigorous approach for the textile micro- and mesomechanical analysis appeared in a later stage. Obstacles for the successful use of numerical methods were mainly the large displacement effects and the nonlinearity related with the deformations of textiles and the convergence problems arose. Munro et al. (Munro et al., 1997a) proposed a new approach for the application of FEM to the aligned fibre assembly problem. Three dimensional 8-node elements with cuboid shape in the neutral configuration and 6 degrees of freedom (DOF) per node employed for the investigation. The approach attempted to separate the various energy contributions to the element stiffness, allowing the user to specify their properties individually. This technique was successful in the easy introduction of nonlinear material properties in the solid model. The approach of Munro et al. (Munro et al., 1997b) was verified qualitatively by modelling realistic yarn situations. The yarn models were meshed by dividing them into layers where the layer interfaces were surfaces perpendicular to the yarn axis. Each layer was split into a number of finite elements ranging from 1 to 22. Initial configurations were arranged so that the fibres within the elements followed idealized helical-yarn geometry. A multi-layer yarn model consisting of 9 elements per layer was subjected to axial extension and axial compression. The model presented the expected, in terms of quality, deformation behaviour. Thus the necking of the yarn piece was caused by the helical winding of the fibres appeared during extension. Moreover the elements of the model were opened significantly during the axial compressing test since the fibres were buckled to avoid compression of the fibre material.

The advance and easy manipulation of CAD tools, in the last few years, allowed the construction of 3D solid models of textile structures. By the use of the attributes of these tools, such as numerical interpolations, mirroring abilities etc. the representation of the structures became feasible. A yarn modelling approach based on the assumprion of helicoid filaments of a constant helix radius and a circular filament cross-section for the loose and a dense structure are presented in the Figure 8 andFigure 9 respectively.

Parametric solid modelling software packages are currently available allowing the construction of complex woven structures (Figure 11). The complete design flexibility provides the selection of weave pattern, yarn size or spacing. The yarn representation is still based on the assumption of the homogenous material for the simplification of modelling and the computational time saving (Toney, 2000). The advance moreover of the FEA codes allowed the mechanical simulation of the unit cells of the modelled textile structures. The mesomechanical modelling of textile structures was improved by the employment of advanced finite elements types and libraries of material properties including linear, nonlinear, elastic, plastic, viscoelastic, isotropic, orthotropic, anisotropic options etc. Additionally the introduction of contact algorithms and large strain effects was essential for the realistic results of the simulated tests. Lin et al. (Lin et al. 2008) studied the mechanical behavior of woven fabrics under compression implementing the finite element analysis using solid elements and nonlinear material properties. Furthermore, Durville (Durville, 2010) approached the textile simulation of woven structures' problem at the fibres scale by means of 3D beam model, providing interesting data useful in the incorporation of fibres in composites structures.

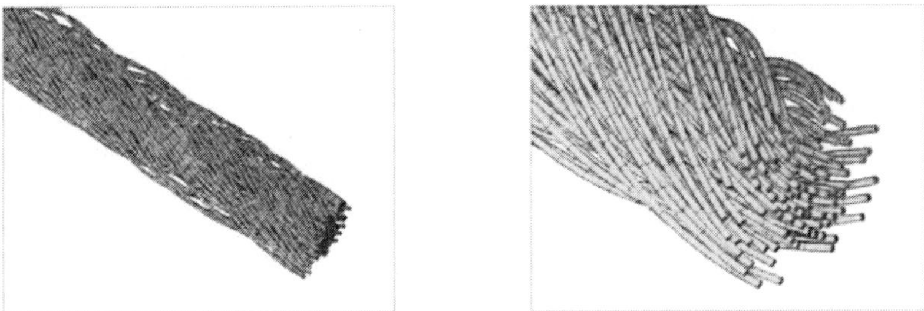

Figure 8. Beam model of filaments in random locations for loose yarn structure.

Significant progress noticed in the modelling of complex structures of fabrics. Tarfaoui and Akesbi (Tarfaoui & Akesbi, 2001) presented the model of the twill woven fabric and the mechanical simulation using the FEM. The unit cell is composed by three warp yarns that intersect with three weft yarns, presenting a different type of crimp. Furthermore, B-spline curve methods have been successfully used to model woven yarns (Turan & Baser, 2010 Jiang & Chen, 2010).

Intensive researches were conducted in the field of woven fabrics composites due to their progressive spread in industrial applications. Actually the exceptional characteristics of woven fabrics composites, as high stiffness and strength, light-weight and efficient manufacturability are determinant for their expansion in automotive, marine and aerospace industry. Zhang and Harding presented one of the first numerical studies for the evaluation of the elastic properties of the plain woven composite structures (Zhang & Harding, 1990). Their approach was based on a strain energy method applied to a one-direction undulation model using the FEM. The drawback of this approach, reported also by the authors, was the consideration of the tow undulation in one-direction that is a non-realistic assumption for woven fabrics. Naik expanded the above approach taking into account the strand cross-section geometry, possible gap between two adjacent strands and the two-direction undulation geometry (Naik & Ganesh, 1992). Actually his detailed model demands a large number of geometrical parameters to describe the undulation and varying thickness of the tow structure. The evolution of numerical methods in the next years produced the first 3D finite element models of the plain woven composites. Whitcomb studied the effect of quadrature order, mesh density and material degradation on the predicted failure resulting from the in-plain loading (Whitcomb & Srirengan, 1996).

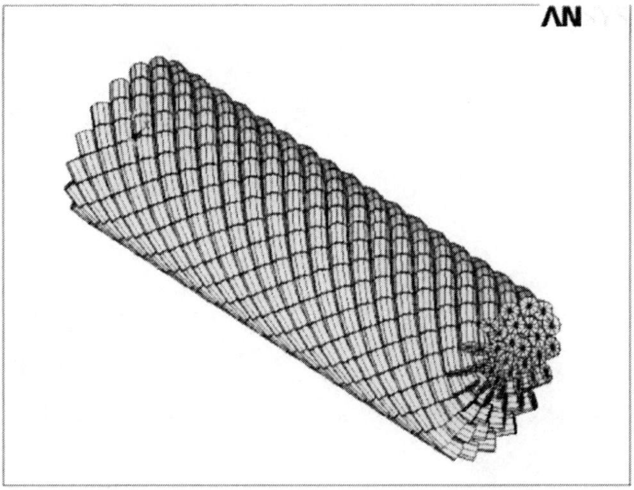

Figure 9. Beam FE model of 50-filament twisted yarn (**Vassiliadis et al., 2010**).

The 3D solid modelling of the composite structure consists in the generation of the volumes representing the woven unit cell and an external volume (with the apparent dimensions of the composite unit cell). Then subtracting volumes of the woven structure from the external volume, the volume of the matrix material is resulted (see figure 12).

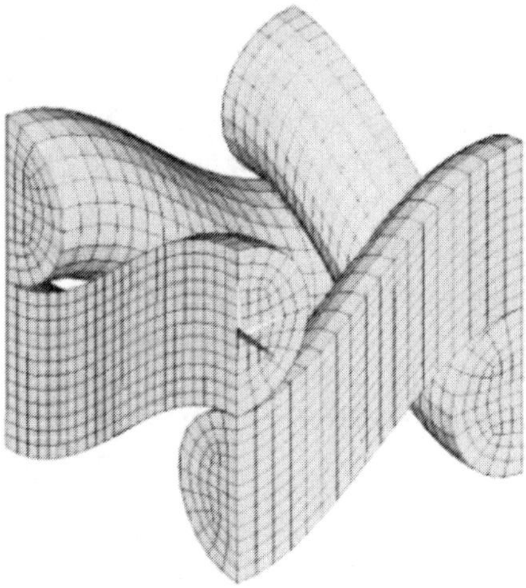

Figure 10. Solid FE model of unit cell of plain woven structure.

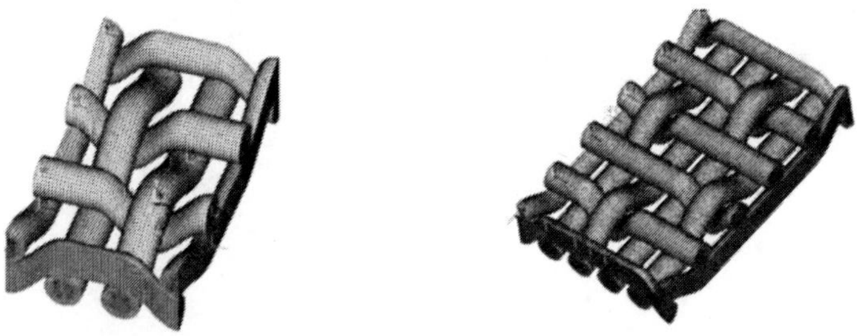

Figure 11. Solid FE model of unit cell of twill (left) and satin (right) woven structure (**Vassiliadis et al., 2008**).

Several approaches were based on the prediction of the homogenized elastic properties of fabric composites using the unit cell of the composite structure. The geometrical representation of the tows was based on certain assumptions such as circular, elliptic, compressed hexagonal and lenticular cross-section areas were considered (Figure 13). The used tows (usually made of glass or carbon fibres) were assumed as transverse isotropic material and the matrix (usually resin) as isotropic material. The homogenized elastic properties of the unit cell results from the mesomechanical analysis using FEM. A relative approach proposed by Ng et al. (Ng et al., 1998) has ben applied for the prediction of the in-plane elastic properties of a single layer 2/2 twill weave fabric composite. The compressed hexagonal shape was considered for the tow cross-section. The modelling and mechanical analysis was programmed using the ANSYS Parametric Design Language (APDL). The 8-node solid elements with 3 degrees of freedom (translational) per node were used. Indicatively a model of approximately 52000 finite elements and 12000 nodes was generated. The contact areas generated during the subtracting operation (for the generation of matrix material) were assigned to be shared entities for both the yarn and the matrix volumes, to ensure the transmission of loading. Choi and Tamma (Choi & Tamma, 2001) dealt with the prediction of the in-plane elastic properties of a composite structure reinforced with plain woven fabric. The predicted elastic properties were used in continue for the damage analysis of the laminated composite structures. The superposition principle was applied for the evaluation of homogenised properties of the woven fabric composite. The generated model of composite unit cells consists of 520 wedge elements for the yarns and 256 brick elements for the matrix. The progressive damage was evaluated simulating the in-plane tensile and shear deformation introducing a respective incrementing load. The degradation of elastic moduli and Poisson ratios was considered for the mechanical damage analysis.

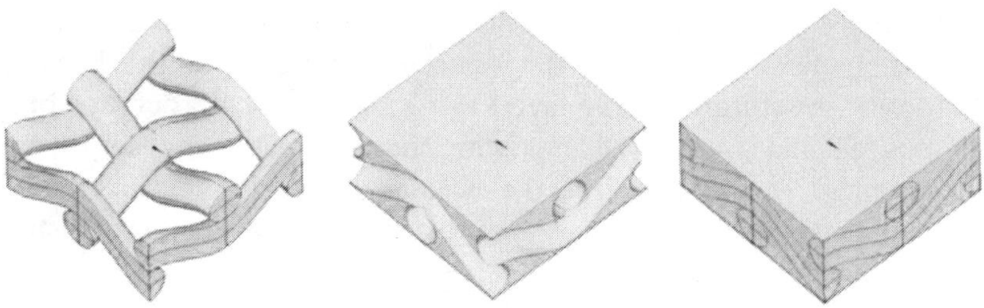

Figure 12. Geometrical model of composite woven structure (woven reinforcement, matrix, composite)

Figure 13. Geometrical model of a woven structure of tows and a composite structure.

A main framework for the multi-scale modelling of woven composite structures for the damage prediction was proposed by Kwon (Kwon, 1993, 2001; Kwon & Hamilton, 1995; Kwon & Roach, 2004) and implemented in several following investigations. It is worth to mention that the damage of a laminated textile composite is presented as a matrix damage, fibre brakeage, fibre – matrix debonding or laminated debonding (delamination). The proposed multi-scale approach is based on the integration of three individual modules: the fibre-strand module, the strand-fabric module and the lamination module. The fibre-strand module aims at the evaluation of the effective elastic properties of a unidirectional composite strand exploiting the material properties and structure of the constituent fibres and matrix. Moreover the current stage relates the stresses and strains of the strand with the stresses and strains of the fibre and matrix materials thus the damage criteria can be applied. The strand-fabric module focuses on the evaluation of the effective properties of the woven fabric composite (unit cell) exploiting the material properties of the unidirectional composite strand. In addition the current stage relates the stresses and strains of the composite structure with the stresses and strains of the strand. Finally the lamination module evaluates the effective properties of the laminated composite structure (multiple layer) using the material properties of the composite lamina. A classical lamination theory or a higher order theory is implemented in this stage. Thus the stresses and strains developed on the laminated composite structure are correlated with the stresses and strains of the lamina.

An innovative research in the field of fabric composites is conducted in the K.U. Leuven, initially focusing on the generalized description of the internal structure of the textile reinforcement. Lomov and his colleagues developed a model for the internal geometry of 2D- and 3D-weaves based on a minimum number of topological data and yarn mechanical properties. The mechanical model applies a yarn deformation energy minimization algorithm to predict the internal geometry of any 2D- and 3D-weave. This approach was systematically extended to 2D- and 3D-woven, two- and three-axial braided, weft knitted and non-crimp warp-knit stitched fabrics and laminates and incorporated in the Wise-Tex software package (Verpoest & Lomov, 2005; Lomov et al., 2000; Lomov et al., 2001). Regarding the damage analysis of the composite structures a three-level hierarchy was proposed: the micro-, meso- and macro-level. The micro-level defines the arrangement of fibres in the representative volume of the impregnated yarn. The meso-level describes the internal structure of the reinforcement and variations of the fibre direction and volume fraction within the yarn. Finally the macro-level defines the 3D geometry of the composite part and the distribution of the reinforcement properties.

5.2. Macromechanical Modelling Of Complex Deformations

The macromechanical modelling of fabrics or cloth modelling, as usually referred, attracted the interest of the textile community in the last decades. Many investigators attempted to approach computationally the macromechanical performance of fabrics for several purposes from the prediction of the drape behaviour of the fabric up to the virtual mode show (Gray, 1998). Depending on the purpose served and the application field different techniques were developed. The basic classification of the developed techniques is divided into computer animation models (graphic models) and the engineering design models. Many numerical techniques including the particle-based model, the deformable node-bar model and the FEM were developed for the engineering design of fabrics. Most of the efforts were focused on the prediction of the drapeability of fabrics.

The used FEM for the drape simulation were based on a variety of element types from simple rods to complex shell elements. Collier (Collier et al., 1991) studied the drape behaviour of fabrics using a nonlinear FEM based on the classical nonlinear plate theory. The fabric was assumed to be two dimensional. It was considered as a linear elastic material with orthotropic anisotropy, where the symmetry lines are aligned in the warp and weft directions. Many corrective actions were assigned the following years by the researchers in the classical finite element techniques in order the realistic performance of fabrics to be approached.

The FEM and flexible thin shell theory was employed by Chen and Govindaraj (Chen & Govindaraj, 1995) to simulate the fabric drape. Their approach provides nonlinear solution since large displacements appear during drape test. Thus the loads are applied incrementally to the system, and at each step, the equilibrium equation system is solved by a Newton-Raphson method. The nonlinearity was handled by calculating the stiffness matrix in each step as a function of the displacement vector. The fabric was considered continuous orthotropic material. A 9-node, doubly curved shell element with 5 DOF per node was used for the simulation.

The simulation of the 3D drape test based on the FEM was also approached by Kang and Yu (Kang & Yu, 1995). The woven fabric was assumed to be an elastic material with orthotropic anisotropy. The fabric was considered as a thin flexible plate under the plane stress condition, and the transverse shear strain was included in the formulation. Since large displacements and large rotations are developed during draping, the drape phenomenon was considered as geometrically nonlinear and respectively the nonlinear analysis was adopted for the simulation. The Green-Lagrangian strains and the second Piolar-Kirchhoff stresses were used for the analysis. The formulation of the FEM was based on a total Lagrangian approach. 4-node quadrilateral elements were used with 5 DOF in each node. In order to avoid the shear locking phenomenon which is commonly observed in the thin plane analysis, a transverse shear strain interpolation method was applied. Almost the same approach was proposed byGan et al. (Gan et al., 1995). In their analysis 8-node shell elements were used with 5 DOF per node. The adopted technique in this approach for the elimination of locking was a reduced integration with zero energy mode control.

For the minimization of the computational power required for the simulation of fabric drape, a FEM using simple beam elements with 6 DOF per node was proposed by Ascough et al. (Ascough et al., 1996). The used beam elements include mass and stiffness properties and can represent iso- or orthotropic cloth properties. The large displacement effects were achieved with the addition of a geometric or initial stress matrix to the elastic stiffness matrix to form the element characteristic matrix. Newmark's method was used to allow a time-stepping approach to the solution, with the advantage that the mesh geometry can be updated at each step. The proposed analysis includes also interaction of the cloth with the body form. Checks for a collision detection of material elements with the body model are made following each time step of the drape simulation. An iterative calculation process is executed until contact rather than penetration of cloth element with the body model occurs.

An approach for the drape simulation of woven fabrics quite different from the traditional macromechanical methods was proposed by Breen et al. (Breen et al., 1994). The cloth was modelled as a collection of particles that conceptually represent the crossing points of warp and wefts threads in a plain weave. Important mechanical interactions that determine the behaviour of woven fabric are discretized and lumped at these crossing points. The various yarn-level structural constraints are represented with energy functions that capture simple geometric relationships between the particles. These energy functions account for the four basic mechanical interactions of yarn collision, yarn stretching, out of plane bending, and trellising. The simulation was implemented as a three-phase process operating over a series of discrete time steps. The first phase for a single time step calculates the dynamics of each particle and accounts the collisions between particles and surrounding geometry. The second phase performs an energy minimization to enforce inter-particle constraints. The third phase corrects the velocity of each particle to account for particle motion during the second phase.

Stylios et al. (Stylios et al., 1995; Stylios et al., 1996) proposed a node-bar model for the drape modelling of fabrics. The deformable elements were defined as consisting of one deformable node with a number of rigid bars. Thus the patch of cloth is divided into a grid (the patch is divided as a series of elements, which can be of equal or unequal sizes). The material properties of the continuum in all elements are lumped together at these deformable nodes by integrating all the energies within those elements. The total energy density was considered as the sum of strain, kinetic energy density, and the energy density introduced by external and boundary forces. Viscoelastic terms were added in the energy equation. The cloth motion in continue was determined using the Euler-Lagrange equations.

The finite volume method employed by Hu et al. (Hu et al., 2000) for the drape modelling of fabrics. The mesh lines were aligned along the warp and weft direction producing rectangular internal volumes and triangular or quadrilateral boundary volumes in a circular fabric sheet. The equilibrium equations of the fabric sheet derived using the principle of stationary total potential energy. Geometric nonlinearity and linear elastic orthotropic material properties of the fabric were considered in the formulation. The full Newton-Raphson iteration method with line searches was adopted for the solution of the resulting nonlinear algebraic equation.

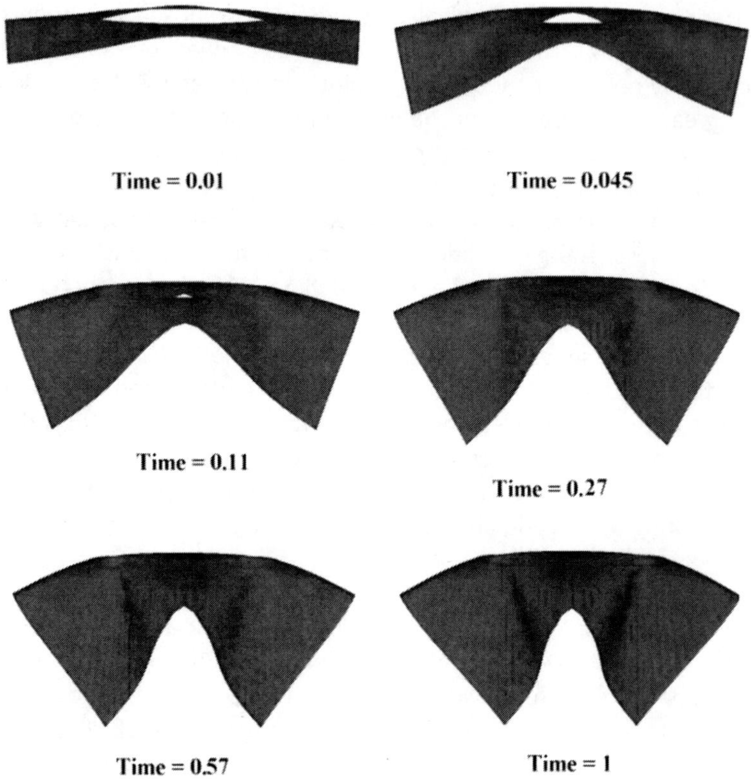

Figure 14. Deformed FE model of square fabric in drape test (**Provatidis et al., 2009**).

5.3. Evaluation Of The Numerical Methods

The adoption of computational techniques in textile mechanics is essential to face and overcome the objective difficulties, as the geometrical representation, the complex deformations, the particular material properties, the contact phenomena and the large deflection effects. Moreover, the advanced computer based tools are suitable for the virtual representation of a product performance under loading. That is a significant facility for the textile designers since a realistic sense from the mechanical up to the aesthetic attributes can be provided.

Most of the mesomechanical modeling approaches implemented the finite element method using solid FE. The yarns were assumed as homogenous material with transverse isotropic elastic properties. The attribution of the yarn properties constitutes basic factor for the accuracy of the mesomechanical modelling stage. Thus the equivalent performance of the homogenous yarn, considering the discrete structure, in the tensile and bending deformation is required at least for the reliable attribution of yarn models. It is remarkable that most of the proposed models omitted the calculation of the real value of the yarn bending rigidity and its attribution at the modelled yarn.

The macromechanical modelling approaches are grouped in two basic categories. The first corresponds to the investigations based on the experimental measurement of the mechanical properties of fabrics and the generation of equivalent models describing their bending performance and drapeability (Collier et al., 1991; Ascough et al., 1996; Stylios et al., 1995; Hu et al., 2000; Araujo et al., 2004). The second category focused on the computational analysis of fabrics in the mesoscopic level and the generation of models presenting equivalent in-plane elastic properties (Ng et al., 1998; Choi & Tamma, 2001; Lomov et al., 2007).

The basic drawback encountered in the existing modelling approaches concerns the collaboration of the different modelling stages (micro, meso, macro) for the development of an integrated design procedure of the textile structures. Thus the modelling of the structure in the mesoscopic level should incorporate the micromechanical performance of the yarns. Whereas the modelling of the structure in the macroscopic level should incorporate the mesoscopic performance of the unit cells and therefore the microscopic performance of the yarns. Consequently the collaboration of the discrete modelling stages is attainable generating realistic models and attributing the equivalent properties.

6. CONCLUSIONS

An extended review was conducted over the textile mechanical modelling area. It is obvious that despite the about 70 years of actual research it's not possible to conclude in an Integrated Computer Aided Engineering Environment. The absence of a global tool was remarked, that aggravates the textile design procedure in terms of time and cost.

The structural hierarchy of the textile structures (fibre – yarn – fabric) is correlated with the high level of complexity presented in the modelling procedure and the mechanical analysis of them. The difficulties are increased due to the high divergence of the dimensions corresponding to the fabric sheet (10^{-1} to 10^{0} m) and the structural elements (fibre diameter, 10^{-5} m). The modelling complexity resulted from the structural hierarchy of textiles is faced adopting a relative modelling hierarchy. Thus three basic modelling scales were developed: the micromechanical modelling of yarns, the mesomechanical modelling of the fabric unit cell and the macromechanical modelling of the fabric sheet. The modular modelling of the textile woven fabrics is a systematic method to overcome the complexity of the mechanical structure and the nature of the materials involved. The global evolution of the modelling approaches seem to converge in this stepwise method and thus indicate a likely way towards the desired Textile Computer Aided Engineering environment.

REFERENCES

1. G. M. Abbott, P. Grossberg, G. A. V. Leaf, 1971 The mechanical properties of woven fabrics, part VII: hysteresis and bending of woven fabrics. Textile Research Journal, 41 4 345 348.

2. G. M. Abbott, P. Grosberg, G. A. V. Leaf, 1973 The elastic resistance to bending of plain-woven fabrics. Journal of the Textile Institute, 64 3 , 346 362.

3. J. Amirbayat, J. W. S. Hearle, 1989 The anatomy of buckling of textile fabrics: Drape and conformability of dimensionless groups. Journal of the Textile Institute, 80 51 70.

4. M. Araujo, R. Fangueiro, H. Hong, 2004 Modelling and simulation of the mechanical behaviour of weft-knitted fabrics for technical applications. Part IV: 3D FEA model with a mesh of tetrahedric elements. Autex Research Journal, 4 2 72 80.

5. J. Ascough, H. E. Bez, A. M. Bricis, 1996 A simple beam element, large displacement model for the finite element simulation of cloth drape. Journal of the Textile Institute, 87 1 152 165.

6. S. Backer, 1952 The mechanics of bent yarns. Textile Research Journal, 22(5), 668 681.

7. B. Behre, 1961 Mechanical properties of textile fabrics, part I: Shearing. Textile Research Journal, 31 2 , 87 99.

8. A. E. Bogdanovich, 2006 Multi-scale modeling, stress and failure analyses of 3-D woven composites. Journal of Materials Science, 41 20 , 6547 6590.

9. D. E. Breen, D. H. House, M. J. Wozny, 1994 Predicting the drape of woven cloth using interacting particles. Computer Graphics, 4 365 372.

10. B. Chen, M. Govindaraj, 1995 Physically based model of fabric drape using flexible shell theory. Textile Research Journal, 65 6 324 330.

11. G. Cho, S. Lee, J. Cho, 2009 Review and reappraisal of smart clothing. International Journal of Human-Computer Interaction, 25 6 582 617.

12. J. Choi, K. K. Tamma, 2001 Woven fabric composites, Part I: Predictions of homogenized elastic properties and micromechanical damage analysis. Int. J. Numer. Meth. Eng,50 2285 2298.

13. K. F. Choi, S. K. Tandon, 2006 An energy model of yarn bending. Journal of the Textile Institute, 97 1 49 56.

14. J. R. Collier, B. J. Collier, G. O'Toole, S. M. Sargand, 1991 Drape prediction by means of finite-element analysis. Journal of the Textile Institute, 82 1 96 107.

15. B. Dahlberg, 1961 Mechanical properties of textile fabrics Part II: Buckling. Textile Research Journal, 31 2 94 99.

16. P. H. Dastoor, T. K. Ghosh, S. K. Batra, S. P. Hersh, 1994 Computer-assisted structural design of industrial woven fabrics part III: modelling of fabric uniaxial/biaxial load-deformation. Journal of the Textile Institute, 85 2 135 137.

17. Dunne, E. L., Brady, S., Smyth, B. and Diamond, D., (2005) Initial development and testing of a novel foam-based pressure sensor for wearable sensing. Journal of NeuroEngineering and Rehabilitation, Vol.2, No.4, pp.1-7.

18. W. D. Freeston, M. M. Platt, M. M. Schoppee, 1967 Mechanics of elastic performance of textile materials. XVIII. Stress-strain response of fabrics under two-demensional loading. Textile Research Journal, 37 11 948 975.

19. W. D. Freeston, M. M. Schoppee, 1975 Geometry of Bent Continuous-Filament Yarns. Textile Research Journal, 45 12 835 852.

20. L. Gan, N. G. Ly, G. P. Steven, 1995 A study of fabric deformation using nonlinear finite elements. Textile Research Journal, 65 11 660 668.

21. S. Gray, 1998 In virtual fashion. IEEE Spectrum, 35 2 18 25.

22. P. Grosberg, 1966 The mechanical properties of woven fabrics, Part II: The bending of woven fabrics. Textile Research Journal, 36 3 205 211.

23. P. Grosberg, S. Kedia, 1966 The Mechanical Properties of Woven Fabrics, Part I: The Initial Load Extension Modulus of Woven Fabrics. Textile Research Journal, 36 1 71 79.

24. R. Haas, A. Dietzius, 1918 The stretching of the fabric and the shape of the envelope in non-rigit ballons. Annual Report, Report 16 National Advisory Committee for Aeronautics, 149 271 (originally published in German as: Haas, Rudolf und Alexander Dietzius, Stoffdehnung und Formaenderung der Huelle bei Prall-Luftschiffen. Untersuchungen im Luftschiffbau der Siemens-Schukkert-Werke, 1913, Luftfahrt und Wissenschaft, Hft. 4.)

25. J. W. S. Hearle, 2006 Engineering design of textiles. Indian Journal of Fibre and Textile Research, 31 1 134 141.

26. J. W. S. Hearle, H. M. A. E. El -Behery, V. M. Thakur, 1959 The mechanics of twisted yarns: Tensile properties of continuous-filament yarns. Journal of the Textile Institute, 50 T83 T111.

27. Horrocks A. R. and Anand, S. C. (Eds.)(2000) Handbook of Technical Textiles, Woodhead Publishing, 978-1-85573-385-5.

28. http://smarteconomy.typepad.com/smart_economy/2006/09/lumalive_a_smar.html

29. http://www.crunchwear.com/cute-circuit-galaxy-led-dress/

30. http://www.ilcdover.com/Space-Suits/

31. http://www2.dupont.com/Kevlar/en_US/index.html

32. http://www.soltis-textiles.com/

33. http://waldemeyer.com/projects/fashion/interactive-led-dance-shoe-disney/

34. J. L. Hu, 2004 Structure and mechanics of woven fabrics, Woodhead Publishing Limited, 1 85573 904 6.

35. J. Hu, S.. Chen, J. G. Teng, 2000 Numerical drape behavior of circular fabric sheets over circular pedestals. Textile Research Journal, 70 7 593 603.

36. J. L. Hu, J. G. Teng, 1996 Computational fabric mechanics: Present status and future trends. Finite Elements in Analysis and Design, 21 4 225 237.

37. N. C. Huang, 1979a Finite Biaxial Extension of Completely Set Plain Woven Fabrics. Journal of Applied Mechanics, Transactions ASME, 46 3 651 655.

38. N. C. Huang, 1979b Finite biaxial extension of partially set plain woven fabrics. International Journal of Solids and Structures, 15 8 615 623.

39. Jayaraman S. Kiekens P. Grancaric A. M. Eds 2006 Intelligent Textiles for Personal Protection and Safety, IOS Press Inc., 1586035991.

40. A. Kallivretaki, 2010 Three-dimensional micromechanical models of textile fabrics, PhD Thesis, National Technical University of Athens.

41. T. J. Kang, W. R. Yu, 1995 Drape simulation of woven fabric by using the finite-element method. Journal of the Textile Institute, 86 4 635 648.

42. S. Kawabata, M. Niwa, H. Kawai, 1973 Finite-Deformation Theory of Plain-Weave Fabrics- 1. The Biaxial-Deformation Theory. Journal of the Textile Institute, 64 1 21 46.

43. M. Keefe, D. C. Edwards, J. Yang, 1992 Solid modeling of yarn and fiber assemblies. Journal of the Textile Institute, 83 2 185 196.

44. A. Kemp, 1958 An extension of Peirce's cloth geometry to the treatment of non-circular threads. Journal of the Textile Institute, 49 1 T44 T48.

45. T. Komori, 2001 A generalized micromechanics of continuous-filament yarns part I: Underlying formalism. Textile Research Journal, 71 10 898 904.

46. M. Konopasek, 1980a Classical elastica theory and its generalizations. In: J.W.S. Hearle, J.J. Thwaites and J. Amirbaya, eds, Mechanics of flexible fibre assemblies, Nato Advanced Study Institutes Series, Series E: Applied Science- 38 USA: Sijthoff & Noordhoff, 255 274.

47. M. Konopasek, 1980b Computational aspects of large deflection analysis of slender bodies. In: J.W.S. Hearle, J.J. Thwaites and J. Amirbaya, eds, Mechanics of flexible fibre assemblies, Nato Advanced Study Institutes Series, Series E: Applied Science- 38 USA: Sijthoff & Noordhoff, 275 292.

48. M. Konopasek, 1980c Textile application of slender body mechanics. In: J.W.S. Hearle, J.J. Thwaites and J. Amirbaya, eds, Mechanics of flexible fibre assemblies, Nato Advanced Study Institutes Series, Series E: Applied Science- 38 USA: Sijthoff & Noordhoff, 293 310.

49. Y. W. Kwon, 1993 Calculation of effective moduli of fibrous composites with micro-mechanical damage. Composite Structures, 25 1-4 , 187 192.

50. Y. W. Kwon, 2001 Multi-level approach for failure in woven fabric composites. Advanced Engineering Materials, 3 9 713 717.

51. Y. W. Kwon, B. H. Hamilton, 1995 Micro/macro-analysis of damage evolution in plate bending of composites. Proceedings of the 1995 ASME International Mechanical Engineering Congress and Exposition; San Francisco, CA, USA; November 1995, 321 1 9.

52. Y. W. Kwon, K. Roach, 2004 Unit-cell model of 2/2-twill woven fabric composites for multi-scale analysis. CMES- Computer Modeling in Engineering and Sciences, 5 1 63 72.

53. J. Lindberg, B. Behre, B. Dahlberg, 1961 Mechanical properties of textile fabrics, Part III: Shearing and buckling of various commercial fabrics. Textile Research Journal, 31 2 99 122.

54. D. W. Lloyd, F. Mete, K. Hussain, 1996 An approach to the theoretical mechanics of static drape. International Journal of Clothing Science and Technology, 8 3 43 58.

55. W. M. Lo, J. L. Hu, L. K. Li, 2002 Modeling a fabric drape profile. Textile Research Journal, 72 5 454 463.

56. S. V. Lomov, A. V. Gusakov, G. Huysmans, A. Prodromou, I. Verpoest, 2000 Textile geometry preprocessor for meso-mechanical models of woven composites. Composites Science and Technology, 60 11 2083 2095.

57. S. V. Lomov, G. Huysmans, Y. Luo, R. S. Parnas, A. Prodromou, I. Verpoest, F. R. Phelan, 2001 Textile composites: Modelling strategies. Composites- Part A: Applied Science and Manufacturing, 32 10 1379 1394.

58. S. V. Lomov, D. S. Ivanov, I. Verpoest, M. Zako, T. Kurashiki, H. Nakai, S. Hirosawa, 2007 Meso-FE modelling of textile composites: Road map, data flow and algorithms. Composites Science and Technology, 67 9 1870 1891.

59. W. A. Munro, G. A. Carnaby, A. J. Carr, P. J. Moss, 1997a Some Textile Applications of Finite-element Analysis. Part I: Finite Elements for Aligned Fibre Assemblies. Journal of the Textile Institute, 88 4 325 338.

60. W. A. Munro, G. A. Carnaby, A. J. Carr, P. J. Moss, 1997b Some Textile Applications of Finite-element Analysis. Part II: Finite Elements for Yarn Mechanics. Journal of the Textile Institute, 88 4 339 351.

61. N. K. Naik, V. K. Ganesh, 1992 Prediction of on-axes elastic properties of plain weave fabric composites. Composites Science and Technology, 45 2 135 152.

62. S.-. Ng, P.. Tse, K.-. Lau, 1998 Numerical and experimental determination of in-plane elastic properties of 2/2 twill weave fabric composites. Composites Part B: Engineering, 29 6 735 744.

63. B. Olofsson, 1964a A general model of a fabric as a geometric-mechanical structure. Journal of the Textile Institute, 55 11 T541 T557.

64. B. Olofsson, 1964b The Setting of Wool Fabrics- A Theoretical Study. Journal of the Textile Institute, 20 272 273.

65. E. Önder, G. Bacer, 1996 A comprehensive stress and breakage analysis of staple fiber yarns Part I: Stress analysis of a staple yarn based on a yarn geometry of conical helix fiber paths. Textile Research Journal, 66 9 562 575.

66. J.. Park, A.-. Oh, 2006 Bending rigidity of yarns. Textile Research Journal, 76 6 478 485.

67. E. M. Parsons, T. Weerasooriya, S. Sarva, S. Socrate, 2010 Impact of woven fabric: Experiments and mesostructure-based continuum-level simulations. Journal of the Mechanics and Physics of Solids, 58 1995 2021.

68. F. T. Peirce, 1937 The geometry of cloth structure. Journal of the Textile Institute, 28 T45 43 77.

69. M. M. Platt, W. G. Klein, W. J. Hamburger, 1959 Mechanics of elastic performance of textiles materials, part XIV: Some aspects of bending rigidity of single yarns. Textile Research Journal, 29 611

70. J. R. Postle, R. Postle, 1996 Modelling fabric deformation as a nonlinear dynamical system using Bäcklund Transformations. International Journal of Clothing Science and Technology, 8 3 22 42.

71. C. Provatidis, S. Vassiliadis, 2002 On the numerical estimation of the mechanical behaviour of fabrics, 2nd AUTEX World Textile Conference, 2002,, 88 97.

72. C. G. Provatidis, S. G. Vassiliadis, 2004 On the performance of the geometrical models of fabrics for use in computational mechanical analysis. International Journal of Clothing Science and Technology, 16 5 434 444.

73. C. G. Provatidis, S. G. Vassiliadis, E. A. Anastasiadou, 2005 Contact mechanics in two-dimensional finite element modelling of fabrics. International Journal of Clothing Science and Technology, 17 1 29 40.

74. C. Provatidis, A. Kallivretaki, S. Vassiliadis, 2009 Fabric Drape Simulation using FEM, Proceedings of the South-East European Conference on Computational Mechanics, SEECCM, Rhodes, Greece, June 2009.

75. W. J. Shanahan, D. W. Lloyd, J. W. S. Hearle, 1978 Characterizing the elastic behaviour of textile fabrics in complex deformation. Textile Research Journal, 48 495 505.

76. D. M. Stump, W. B. Fraser, 1996 A simpified model of fabric drape based on ring theory. Textile Research Journal, 66 8 506 514.

77. G. Stylios, T. R. Wan, N. J. Powell, 1995 Modeling the dynamic drape of fabrics on synthetic humans, a physical, lumped-parameter model. Int. J. Clothing Sci. Technol., 7 5 10 25.

78. G. K. Stylios, T. R. Wan, N. J. Powell, 1996 Modelling the dynamic drape of garments on synthetic humans in a virtual fashion show. International Journal of Clothing Science and Technology, 8 3 95 112.

79. N. Takano, Y. Uetsuji, Y. Kashiwagi, M. Zako, 1999 Hierarchical modelling of textile composite materials and structures by the homogenization method. Modelling and Simulation in Materials Science and Engineering, 7 2 207 231.

80. S. L. P. Tang, 2007 Recent developments in flexible electronics for monitoring applications. Transactions of the Institute of Measurement and Control, 29 283 300.

81. M. Tarfaoui, S. Akesbi, 2001 Numerical study of the mechanical behaviour of textile structures. International Journal of Clothing Science and Technology, 13 3-4, 166 175.

82. M. M. Toney, 2000 Computer modeling of fibrous structures. Journal of the Textile Institute, 91 3 133 139.

83. L. A. Tognetti, N. Carbonaro, G. Zupone, D. De Rossi, 2006 Characterization of novel data glove based on textile integrated sensors. Proceedings of the 28th IEEE, EMBS Annual International Conference, 2510 2513.

84. S. Vassiliadis, A. Kallivretaki, A. M. Grancaric, S. Giannakis, C. Provatidis, 2008 Computational modelling of twill and satin woven structure, Proc. of the World Conference AUTEX 2008, June 2008, Biella, Italy.

85. S. Vassiliadis, A. Kallivretaki, C. Provatidis, 2010 Mechanical modelling of multifilament twisted yarns. Fibers and Polymers, 11 1 89 96.

86. I. Verpoest, S. V. Lomov, 2005 Virtual textile composites software WiseTex: Integration with micro-mechanical, permeability and structural analysis. Composites Science and Technology, 65 15-16 SPEC. ISS.), 2563 2574.

87. J. Whitcomb, K. Srirengan, 1996 Effect of various approximations on predicted progressive failure in plain weave composites. Composite Structures, 34 1 13 20.

88. P. J. Xu, H. Zhang, X. M. Tao, 2008 Textile-structured electrodes for electrocardiogram. Textile Progress, 40 4 183 213.

89. Y. C. Zhang, J. Harding, 1990 A numerical micromechanics analysis of the mechanical properties of a plain weave composite. Computers & Structures, 36 5 839 844.

Chapter 5

Liquid Transport in Nylon 6.6. Woven Fabrics Used for Outdoor Performance Clothing

A. B. Nyoni[1]

[1] National University of Science and Technology, Department of Textile Technology, Zimbabwe

1. INTRODUCTION

Humans rely on the evaporation of sweat to remain comfortable and prevent overheating in hot environments and during exercise.[1] Discomfort results from the build up of sweat on the skin and if it doesn't evaporate quickly, the body core temperature heats up producing more sweat exposing the wearer to potential afflictions such as post-exercise chill and even hypothermia. Therefore, with properly engineered dynamic or responsive fabrics [2,3] less energy to cool the body will be required resulting in increased performance and endurance.

Researchers [4,5] generally agree that liquid transport properties are significantly affected by fibre type, yarn construction and fabric construction. The fibre length, width, shape and alignment all have a great influence on the quality of the capillary channels in the inter-fibre spaces and size of the pores present. The density and structure of yarns can greatly influence the dimensions and structure of inter- and intra-yarn pores[4] and pore sizes and distribution are determined by the manner in which fibres are assembled into the woven, nonwoven, or knitted structure.[6] Finishing treatment of the fabric surface and its surface roughness and the bulk properties of the liquid (i.e. viscosity, surface tension, volatility and stability) also play a significant role during wicking.

Additional important variables which exert influence on wicking are the level of physical activity and environmental conditions such as the relative humidity of the atmosphere which combined with the ambient temperature, determine the water vapour pressure of the ambient atmosphere and hence the rate of water vapour transfer through clothing. The wind speed which affects the thermal and water vapour resistance of the air adjacent to the fabric also plays a significant part during wicking.[7]Therefore, to design textile materials with specific functional properties of moisture management, it is essential to establish the relationship between the wicking properties of yarns and the structure of the fabric they are part of. In this chapter the effect of these variables on the wicking performance of a selected fabrics made from a combination of textured and flat continuous Nylon 6.6 yarns[8] were determined by The Longitudinal Wicking "Strip" Test using BS3424 Method 21 (1973).

In all the fabrics, saturated, unsaturated and dry zones were exhibited and the simultaneously occurrence of wetting, wicking, liquid dispersion and evaporation influenced the time exponent values*k* obtained.

The critical volume of liquid at which transfer wicking occurred at yarn cross over regions termed as the "transfer rate" was influenced by two competitive effects, i.e. the tendency to spread in the capillary space between the filaments of "absorber" textured yarns and the tendency to wick the liquid by the "runner"flat continuous filament yarns yarns

2. FABRIC SAMPLE PREPARATION AND TEST METHODS

Fabrics woven from different combinations of nylon 6.6 filament yarns were selected and the characteristics determined as shown in Table 1.

Prior to testing, the samples were conditioned in a standard atmosphere of 20±2ºC and 65±2% relative humidity for 24 hours. Sample strips of 3.5cm x 33cm each were cut in the warp and weft directions from the conditioned sample. To aid observation of the wicking distance, a pen filled with water soluble ink was used to mark a graduated scale in 1cm intervals on the strips. The samples were then mounted on the pinned frame for the vertical, horizontal and syphon tests as shown in Figures 1, 2 and 7 respectively. The dipping ends of the samples were aligned leaving a length of 1cm to dip into the infinite reservoir containing distilled water. A ruler with millimeter divisions was placed parallel to the sample strip to enhance the accuracy of the measurement.

For washed fabric tests, the fabric samples were washed with a non-biological detergent in an automatic front loading domestic washing machine and tumble dried according to the ISO 6330:2000 which specifies domestic washing and drying procedures for textile testing. The dry fabrics were then conditioned in a standard atmosphere of 20± 2 C and 65±2% relative humidity for 24 hours before testing. Sample strips of 3.5cm x 33cm each were cut in the warp and weft directions from the conditioned fabric sample and tested with the frame in both the vertical and horizontal positions above the water basin containing distilled water and the results are shown in Table 2.

Figure 1. Vertical Strip Wicking Test

The height of the advancing liquid front as a function of time was recorded by visual observation of the running ink through a travelling microscope at 5 minutes intervals for the first hour, and then at hourly intervals thereafter until the maximum wicking height (equilibrium point) was reached. To avoid contamination by the indicating ink the test liquid was changed after each test. Constant temperature and humidity in the ambient atmosphere were achieved by testing in the conditioned room.

The strip method has been used by Hollmark and Peek[9] to characterize the wicking behaviour of porous materials and they found it readily applicable under different conditions with a relatively high degree of reproducibility. Zhuang[10] also found good correlation between results obtained by manual and automatic testing.

3. VERTICAL STRIP WICKING TEST RESULTS

3.1. Fabric Sample S1f–Unwashed

The results obtained from the wicking tests are shown in Table 2 and in Figures 3. Figure 3 shows that there was rapid wicking for the first 5-10 minutes in both the warp and weft directions and then a significant decrease to a slow rate with the lapse of time until it was difficult to note the level of liquid rise in 5 minutes time intervals. Observations done at hourly intervals thereafter tabulated in Table 2indicate that from 60-180 minutes the fabrics continued wicking at a slow rate until an equilibrium point was reached.

Multiple comparison between means of the actual liquid advancement in the first 15minutes (1stQuarter) Table 3 and the second 15 minutes (2nd Quarter) Table 4 of the hourly test shown in Tables 5indicate that that there was a significant difference in the distance moved by the liquid in both the warp and weft direction wicking with the lapse of time. Wicking in the weft direction was more rapid than in the warp direction and multiple comparison of the actual liquid advancement in the first 15minutes (1st Quarter) of the hourly test in Tables 5 show that there was a significant difference in warp and weft direction wicking. Microscopic examination of fabrics during wicking exhibited an almost linear leading edge in the weft direction and a spiked pattern in the warp direction.

Table 1. Fabric and Yarn Characteristics

Property	Test Method	Sample S1F	Sample S2F
Ends/cm	BS 2862:1984	70	70
Picks/cm	BS 2862:1984	30	50
Linear Density Warp (dtex)	BS 946:1970	44dtexf34 flat fully dull PA 6.6	44dtexf34 flat fully dull PA 6.6
Linear Density Weft (dtex)	BS 946:1970	195dtexf170 Airjet Textured Bright PA6.6	44dtexf34 flat fully dull PA 6.6
Fabric Weight g/m^2	BS 2471:1978	43.75	26.31
Filaments x-section	Microscopy-SEM	Circular	Circular
Warp FilamentO Weft	Microscopy-SEM	11.673μm 11.673μm	11.673μm 11.673μm

3.2. Vertical Wicking Fabric Sample S2f-Unwashed

The results in Table 2 and Figures 4 to 5 show that there was rapid wicking for the first 5-10minutes in both the warp and weft directions which became less rapid with the lapse of time. Multiple comparison of wicking results in Table 6 show a significant decrease in weft direction wicking compared to warp direction wicking. The wicking rate significantly decreased to a slow rate with the lapse of time in the warp and weft directions. The rapid attainment of the equilibrium point when wicking the fabrics in the warp and weft direction indicates that the liquid is rapidly spread over a large area for quick evaporation.

4. HORIZONTAL STRIP WICKING TESTS

Wicking occurs when a fabric is completely or partially immersed in a liquid or in contact with a limited amount of liquid such as a drop placed on the fabric. In a vertically held substrate, wicking is affected by gravitational forces and ceases when capillary forces are balanced by the hydrostatic head.[11] At that point, the capillary pressure that raises the liquid is balanced by the effect of gravity, that is, by the weight of raised liquid.[12] To determine the extent to which gravity affects wicking, horizontal wicking tests were carried out on nylon 6.6 fabrics samples S1F and S2F and the results are shown in Table 2.

Table 2. Note: Figures in parentheses indicate the actual liquid advancement per time intervalKey: Uw-UnwashedW-WashedFabric Vertical and Horizontal Wicking Test Results

Wicking time (t)min	Vertical Wicking								Horizontal Wicking							
	S1F-warp (lmm)		S1F-Weft (lmm)		S2F-Warp (lmm)		S2F-Weft (lmm)		S1F-warp (lmm)		S1F-Weft (lmm)		S2F-Warp (lmm)		S2F-Weft (lmm)	
	Uw	W	Uw	W	Uw	W	Uw	W	Uw	W	Uw	W	Uw	W	Uw	W
5	15	35	40	70	28	37	18	25	25	34	50	73	27	32	9	23
10	30(15)	45(10)	56(16)	100(30)	39(11)	45(8)	23(5)	32(7)	33(8)	49(15)	65(15)	101(29)	32(5)	38(6)	13(4)	30(7)
15	35(5)	55(10)	70(14)	119(19)	41(2)	49(4)	27(4)	39(7)	40(7)	55(6)	80(15)	120(19)	36(4)	41(3)	16(3)	32(2)
20	38(3)	62(7)	80(10)	130(11)	44(3)	50(1)	30(3)	41(2)	43(3)	60(5)	90(10)	135(15)	38(2)	43(2)	18(2)	32(0)
25	40(2)	67(5)	90(10)	140(10)	46(2)	51(1)	31(1)	43(2)	46(3)	66(6)	95(5)	149(14)	39(1)	47(4)	19(1)	36(4)
30	45(5)	70(3)	95(5)	146(6)	49(3)	53(2)	31(0)	44(1)	50(4)	71(5)	100(5)	157(8)	40(1)	49(2)	19(0)	38(2)
35	47(2)	72(2)	100(5)	153(7)	49(0)	54(1)	31(0)	44(0)	51(1)	74(3)	103(3)	166(9)	41(1)	49(0)	19(0)	38(2)
40	48(1)	73(1)	104(4)	157(4)	49(0)	55(1)	31(0)	44(0)	52(1)	75(1)	108(5)	169(3)	41(0)	50(1)	19(0)	39(1)
45	49(1)	74(1)	107(3)	158(1)	49(0)	55(0)	32(1)	44(0)	54(2)	78(3)	109(1)	176(7)	41(0)	52(2)	19(0)	39(0)
50	50(1)	75(1)	110(3)	158(0)	49(0)	55(0)	32(0)	44(0)	55(1)	81(3)	111(2)	182(6)	42(1)	52(0)	20(1)	39(0)
55	51(1)	79(4)	112(2)	163(5)	49(0)	55(0)	32(0)	44(0)	56(1)	81(0)	113(2)	185(3)	42(0)	52(0)	20(0)	40(1)
60	52(1)	80(1)	113(1)	165(2)	49(0)	55(0)	32(0)	44(0)	57(1)	84(3)	114(1)	188(3)	42(0)	50(0)	20(0)	40(0)
120	56(4)	89(9)	121(8)	180(15)	50(1)	55(0)	32(0)	48(4)	58(1)	110(26)	125(11)	220(32)	47(5)	61(11)	23(3)	49(9)
180	56(0)	89(0)	121(0)	180(0)	50(0)	55(0)	32(0)	48(0)	58(0)	110(0)	129(4)	227(7)	50(3)	61(0)	23(0)	49(0)

Note: Figures in parentheses indicate the actual liquid advancement per time interval

Key: Uw-Unwashed

W-Washed

Table 3. Key: Uw-UnwashedW-WashedFabric Wicking Test 1st Quarter (15 minutes).

Sample	Vertical Wicking								Horizontal Wicking							
	S1F-warp (lmm)		S1F-Weft (lmm)		S2F-Warp (lmm)		S2F-Weft (lmm)		S1F-warp (lmm)		S1F-Weft (lmm)		S2F-Warp (lmm)		S2F-Weft (lmm)	
	Uw	**W**	**Uw**	**W**	**Uw**	**W**	**Uw**	**W**	**Uw**	**W**	**Uw**	**W**	**Uw**	**W**	**Uw**	**W**
1	35	55	70	119	41	49	27	39	40	55	80	120	36	41	16	32
2	33	54	72	120	40.5	50	28	39	40.5	55	82.5	126	37.5	39.5	15.5	31.5
3	34.5	53.5	69	123	39	51	25	40	40	55	81	123	36	42	16.5	32
4	33	55.5	70.5	115.5	42	48	25.5	38	39	54	78	120	38	41.5	17	33
5	34	54	69	118.5	42	48	29	37.5	41	55	84	120	36	42	17	31
Mean	33.9	54.4	70.1	119.2	40.9	49.2	26.9	38.7	40.1	54.8	81.1	121.8	36.7	41.2	16.4	31.9
SD	2.6	3.3	3.74	4.88	2.86	3.14	2.32	2.78	2.83	3.31	4.03	4.94	2.71	2.87	1.81	2.53
SE	1.3	1.65	1.87	2.44	1.43	1.57	1.16	1.39	1.42	1.66	2.01	2.47	1.36	1.44	0.91	1.26
CV	7.68	6.06	5.34	4.1	4.1	6.99	8.62	7.19	7.06	6.04	4.97	4.05	7.38	6.97	11.04	7.92

Key: Uw-Unwashed
W-Washed

Table 4. Key: Uw-UnwashedW-WashedFabric Wicking Test 2nd Quarter (30 minutes).

Sample	Vertical Wicking								Horizontal Wicking							
	S1F-warp (lmm)		S1F-Weft (lmm)		S2F-Warp (lmm)		S2F-Weft (lmm)		S1F-warp (lmm)		S1F-Weft (lmm)		S2F-Warp (lmm)		S2F-Weft (lmm)	
	Uw	W	Uw	W	Uw	W	Uw	W	Uw	W	Uw	W	Uw	W	Uw	W
1	10	15	20	27	8.0	4	4	5	10	16	25	37	4	8	3	6
2	9.5	15	19	25.5	6.5	4	3	3.5	11	16	25	36	4	7.5	4	6.5
3	10	15	20	25	8.0	4.5	3	4.5	11	17	26	37.5	4.5	6.5	2.5	6
4	9.5	14.5	18.5	29	7.5	4.5	4	5.5	11	16.5	24.5	37	4	7.5	3	6
5	9	15.5	18	27.5	7.5	4.5	3	5	10.5	15	25.5	36	4	8.0	3.5	6
Mean	9.6	15	19.1	26.8	7.5	4.3	3.4	4.7	10.7	16.1	25.2	2.71	4.1	7.5	3.2	6.1
SD	1.39	1.73	1.96	2.32	1.23	0.93	0.83	0.97	1.46	1.79	2.25	1.36	0.91	1.23	0.8	1.11
SE	0.69	0.87	0.98	1.16	0.61	0.46	0.41	0.49	0.73	0.9	1.12	7.38	0.45	0.61	0.4	0.55
CV	14.43	11.55	10.23	8.64	16.33	21.57	24.25	20.63	13.67	11.15	8.91	12.78	22.09	16.33	25	18.11

Key: Uw-Unwashed
W-Washed

Table 5. Multiple Comparison Between Wicking Means of Fabric S1F

Fabric samples	Significance	Difference
V15min-warp-uw Vs.		
V15min-weft-uw	0.000↑	***
H15 min-warp-uw	0.000↑	***
V15min-warp-w	0.000↑	***
V30 min-warp-uw	0.000↓	***
V15min-weft-uw Vs.		
H15 min-weft-uw	0.000↑	***
V15min-weft-w	0.000↑	***
V30 min-weft-uw	0.000↓	***
H15min-warp-uw Vs.		
H15min-weft-uw	0.000↑	***
H15 min-warp-w	0.002↑	***
H30min-warp-uw	0.000↓	***
H15min-weft-uw Vs.		
H15 min-weft-w	0.000↑	***
H30min-weft-uw	0.000↓	***

W = Washed fabric W = Unwashed fabric
H = Horizontal wicking V = Vertical Wicking
↑= Wicking decrease ↓= Wicking increase
Significance of differences of fabric wicking:
*****P≤ 0.001,**P≤0.01,*P≤0.05 and Not significant (ns) at P>0.05.**

Table 6. Multiple Comparison Between Wicking Means of Fabric S2F

Fabric samples	Significance	Difference
V15min-warp-uw Vs.		
V15min-weft-uw	0.000↓	***
H15 min-warp-uw	0.000↓	***
V15min-warp-w	0.000↑	***
V30 min-warp-uw	0.000↓	***
V15min-weft-uw Vs.		
H15 min-weft-uw	0.000↓	***
V15min-weft-w	0.000↑	***
V30 min-weft-uw	0.000↓	***
H15min-warp-uw Vs.		
H15min-weft-uw	0.000↓	***
H15 min-warp-w	0.000↑	***
H30min-warp-uw	0.000↓	***
H15min-weft-uw Vs.		
H15 min-weft-w	0.000↑	***
H30min-weft-uw	0.000↓	***

W = Washed fabric UW = Unwashed fabric
H = Horizontal wicking V = Vertical Wicking
↑= Wicking decrease ↓= Wicking increase
Significance of differences of fabric wicking:
*****P≤ 0.001,**P≤0.01,*P≤0.05 and Not significant (ns) at P>0.05.**

4.1. Horizontal Strip Wicking Test Sample S1f-Unwashed Fabric

The results in Table 2 and Figures 3 to 6 exhibited a similar wicking trend as fabrics wicked in the vertical direction in which wicking in the weft direction was more rapid than in the warp direction. However, even though the trend was similar, there was a significant difference in the distance travelled by the wicked liquid compared to vertically wicking in both the warp and weft directions as shown by the results of multiple comparison of the actual liquid wicked during the 1st and 2nd quarters of an hourly test in Table 5. As was the case with vertical wicking, there was rapid wicking for the first 5-10 minutes in the warp and weft directions.

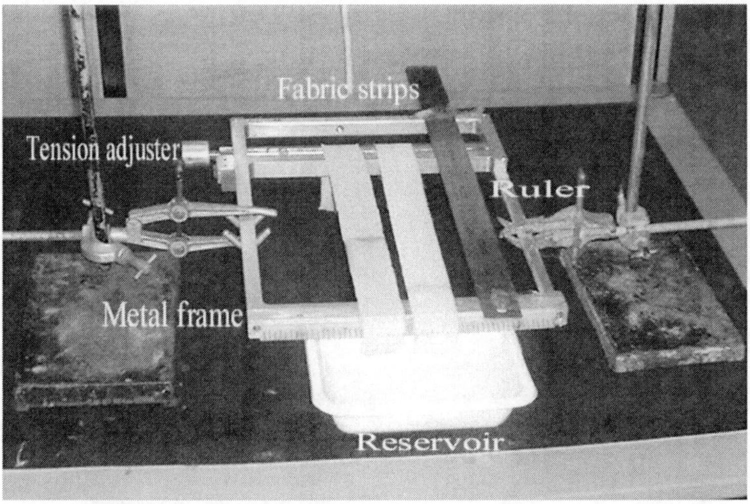

Figure 2. Horizontal Wicking Test

4.2. Horizontal Strip Wicking Test-Sample S2f-Unwashed Fabric

Table 2 and Figures 4 and 5 shows that there was rapid wicking for the first 5-10 minutes but wicking in the warp direction was more rapid than wicking in the weft direction. At the start of wicking there is a variation in lift off followed by the same wicking trend in both the weft and warp directions. Results of multiple comparison of the actual liquid wicked within the 1st and 2nd quarters of an hourly test in Table 5 show a significant decrease in the liquid wicked in both the warp and weft horizontal directions.

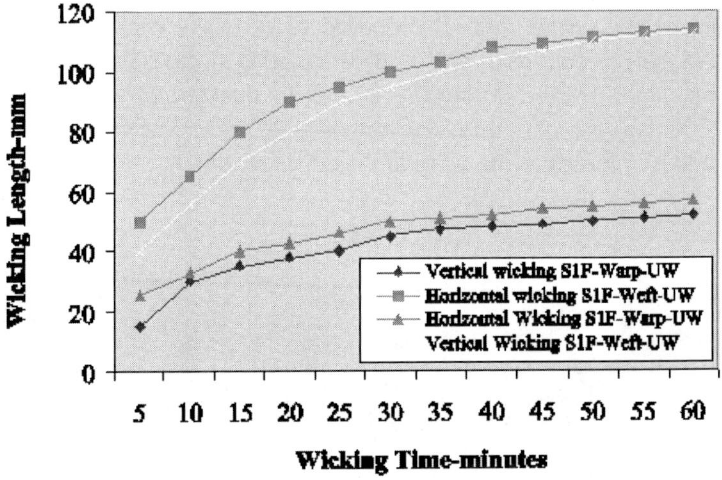

Figure 3. Wicking test of fabric S1F – unwashed fabric

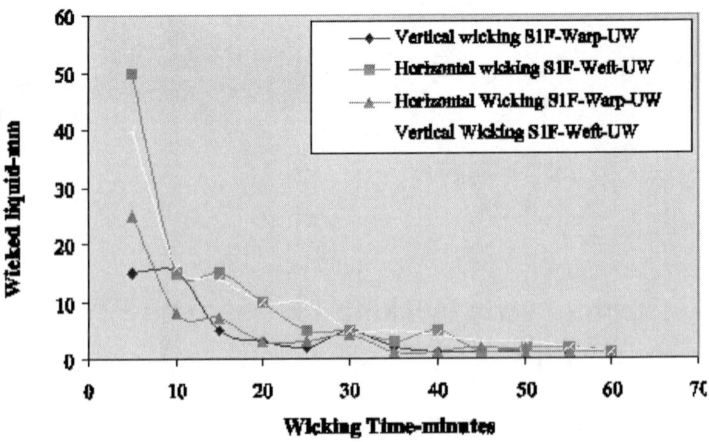

Figure 4. Actual Liquid Advance Sample S1F - Unwashed Fabric

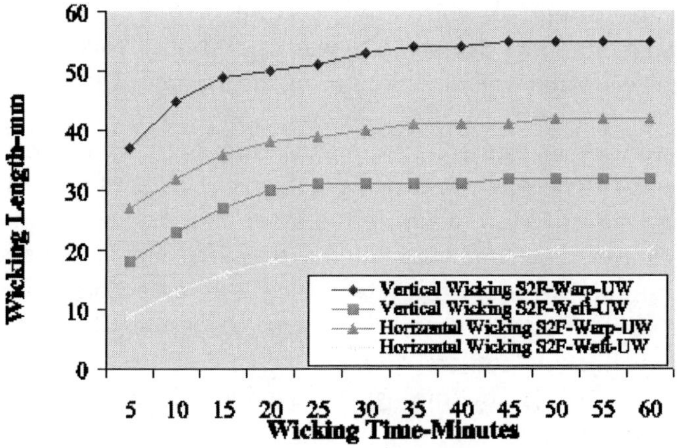

Figure 5. Wicking Tests of Fabric S2F-Unwashed Fabric

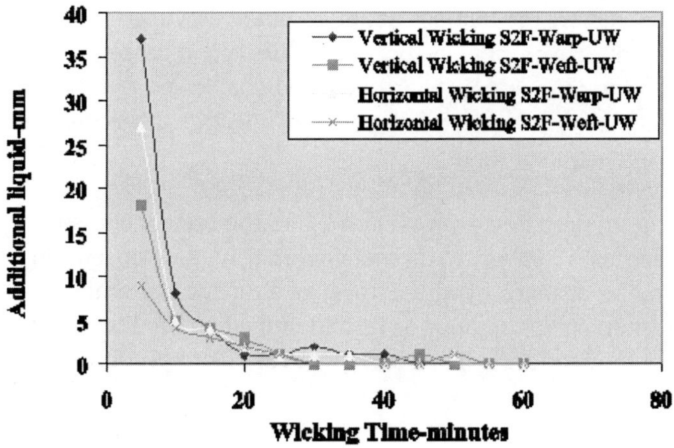

Figure 6. Actual Liquid Advance S2F-Unwashed Fabric

5. SYPHON WICKING

It is a known fact that the liquid flow in downward wicking is aided by gravity and occurs more rapidly through an already saturated fabric with a lower resistance to flow than an initially dry fabric.[13] A further study to determine the extent to which the structure of the constituent yarns affects wicking in fabrics S1F and S2F was carried out by wicking washed fabrics in the warp and weft directions using the Syphon[11] Test Method.

In downward wicking, Figure 7 a rectangular strip of the test fabric is used as a syphon, by immersing one end in a reservoir of water or saline solution and allowing the liquid to drain from the other end at a lower level, into a collecting beaker. The amount of liquid transferred at successive time intervals can be determined by weighing the collecting beaker. No published standards exist and evaluation of results differ between researchers with some authors [14] taking the rate of mass transfer of the liquid when a constant flow through the syphon has been attained as an indicator of wickability.

Hardman[14] distinguished this as a "rate of drainage," using the elapsed time between the initial moment of contact between the fabric strip and liquid and the moment when dripping from the lower fabric end commences as a measure of wicking.

Because of the limited amount of liquid retained by the fabrics S1F and S2F due to the effects of rapid evaporation observed in preceding experiments, determination of their downward wicking behaviour was done by observing the actual distance traveled by the liquid towards the bottom end of the fabric as a function of time.

Samples were prepared as in section 2 and the rectangular strip of the test fabric used as a syphon by immersing 1cm of the top end in the liquid reservoir. The distance of water travel as a function of time was taken at 5 minutes intervals for an hour or terminated when the liquid dripped at the bottom of the fabric or when wicking ceased due to evaporation.

5.1. Syphon Wicking Test- Fabric Sample S1f And S2f

The results in Tables 7 show the distance travelled by the liquid leading front and the figures in parentheses indicate the actual liquid advancement per time interval.

Fabric sample S1F made from 195f170 weft yarn and 44f34 warp yarn with 70 ends/cm and 30 picks/cm (43.75g/m^2) was wicked in the warp and weft directions. Figure 8 shows that after wicking fabric sample S1F for 50 minutes in the weft direction, the liquid had travelled to the lower end of the fabric strip whereas in the warp direction the leading head was still 202mm from the lower end of the fabric.

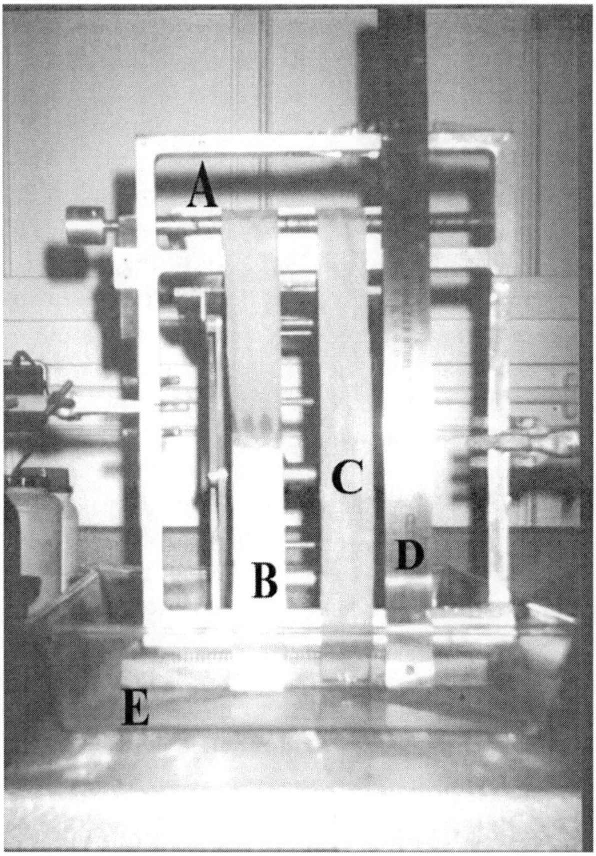

Figure 7. A-liquid reservoirB/C Strips of FabricsD-RulerE-Collecting traySyphon Wicking Test

When the fabric is wicked in the warp direction, the textured weft yarns cause retardation of the liquid's progress due to their absorption capacity. The absorption of the liquid into the heterogeneous structure of the yarn causes a temporary slowing down of its advancement as it is dispersed in the yarn structure before a critical volume is achieved [15] to enable liquid transfer to the capillaries of the warp yarns. The nature of the liquid flow in the warp direction therefore is in fast-slow fast (warp-weft-warp) steps resulting in a haphazard flow as shown in Figure 9. In wicking the fabric in the weft direction, the high volume textured weft yarns rapidly flood the capillaries of the flat continuous filament yarns and this speeds the rate of wicking. The actual advancement of the wicked liquid shown in Figure 9 is directly proportional to the wicking time in both cases (warp and weft) but was found to be 61% more in the weft compared to warp direction wicking. This indicates that for this fabric, the wicking rate does not only depend on the yarn and fabric structure but also on the direction of orientation of the constituent yarns in the structure.

Results in Table 7 show the wicking behaviour of an almost balanced fabric sample S2F made from 44f34 warp and weft flat continuous filament yarns with 70 ends/cm and 50 picks/cm (26.31g/m2g/m2). The graphical representation in Figures 10-11 plotted from the results tabulated in Table 7 show that the rate of warp and weft wicking follow a similar trend. The difference of the actual liquid wicked was 15% more in the warp direction due to the high number of ends/cm compared to picks/cm therefore the packing of the additional filaments in the warp yarns introduced more capillary spaces between the nylon filaments. Due to its light-weight (26.31g/m2g/m2), the fabric allowed rapid liquid evaporation. Results in Table 7 show that the wicking rate had significantly slowed down after 20 minutes despite the fact that the liquid flow in this test was through an already saturated fabric with a lower resistance to flow and was also aided by gravity. After 35minutes wicking, liquid advancement had ceased and when the fabric was left to wick to the end of the hour there was no change in the position of the liquid edge. In the absence of gravity, this indicates that there is significant rapid evaporation of liquid from the fabric which is a desired functional property of fabrics designed to rapidly transmit perspiration to the exterior where it can evaporate.

6. WICKING CHARACTERISTICS OF WASHED FABRICS

The ability of a fibre to facilitate migration of liquid water or water vapour molecules depends on its surface hydrophilicity or affinity for water, the textile finish applied and the fibre substrate.[7,16] The surface properties of man-made fibres are generally adjusted with spin finishing agents during the fibre spinning process.[17] Hydrophobic fibres can be modified in finishing to give surface properties which can allow liquid flow.[18] Leijala and Hautojarvi[17] using Scanning Force Microscopy (SFM) studied the structure, distribution, and composition of spin finish layers on a polypropylene fibre surface. They noted that the coverage and homogeneous distribution of the finish on the fibre surface even though only a few nanometers in thickness is an important factor affecting tribological and antistatic properties as well as the wettability of fibres. In another study[14], the wickability attributed to the conventional (non porous) acrylic as was found to be the case with polypropylene was due to spin finish which could be easily removed by washing. Gogalla[4] also noted that the uneven distribution of chemical finish on the surface of a fabric greatly affected its wicking behaviour.

Electron micrographs of yarns from which the fabric samples S1F and S2F were woven in Figure 12 a-c show traces of spin finish on all the yarns which was removed during the scouring process.

Table 7. Washed Fabric Wicking Tests-Vertical Vs. Syphon Wicking

Sample	S1F-Warp direction (l-mm)		S1F-Weft direction (l-mm)		S2F-Warp direction (l-mm)		S2F-Weft direction (l-mm)	
Wicking Time-t minutes	Vertical	Syphon	Vertical	Syphon	Vertical	Syphon	Vertical	Syphon
5	35	35	70	42	37	58	25	48
10	45 (10)	47(12)	100(30)	84(42)	45 (8)	69(11)	32 (7)	56(8)
15	55 (10)	60(13)	119(19)	129(45)	49 (4)	74(5)	39 (7)	62(6)
20	62 (7)	69(9)	130(11)	156(27)	50 (1)	79(5)	41 (2)	67(5)
25	67 (5)	78(9)	140(10)	199(43)	51 (1)	81(2)	43 (2)	70(3)
30	70 (3)	98(20)	146 (6)	220(11)	53 (2)	82(1)	44 (1)	71(1)
35	72 (2)	110(12)	153 (7)	253(33)	54 (1)	84(2)	44 (0)	73(2)
40	73 (1)	116 (6)	157 (4)	268(15)	55(1)	84(0)	-	-

45	74 (1)	122(6)	158 (1)	282(14)	55(0)	-	-	-
50	75 (1)	128(6)	158 (0)	330(48)	-	-	-	-
55	79 (4)	139(11)	163 (5)	-	-	-	-	-
60	80 (1)	150(11)	165 (2)	-	-	-	-	-
120	89 (9)	-	180(15)	-	-	-	-	-
180	89 (0)	-	180 (0)	-	-	-	-	-

Note:Figures in parentheses indicate the actual liquid advancement per time interval.

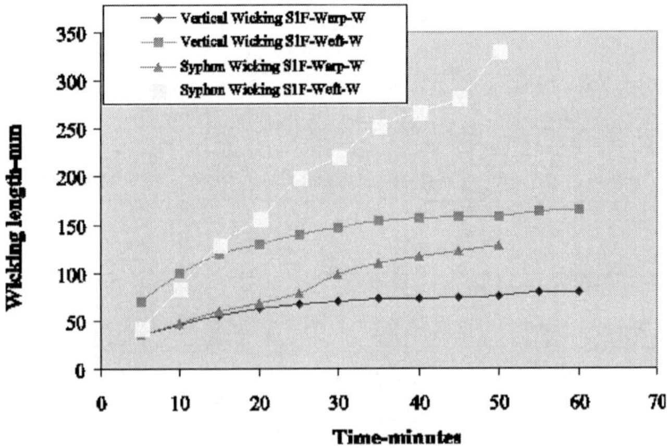

Figure 8. Vertical Vs Syphon Wicking Fabric S1F-Washed

During use, out-door and performance textiles fabrics are exposed to soiling which comes from two different sources, namely,

a) from the body of the wearer and

b) from the environment.

therefore, it will be necessary at some stage to wash the fabrics. However, it is important that such a treatment does not alter the functional properties of these garments. Therefore, it was of interest to study the effect of laundering on the wicking behaviour of fabric samples S1F and S2F.

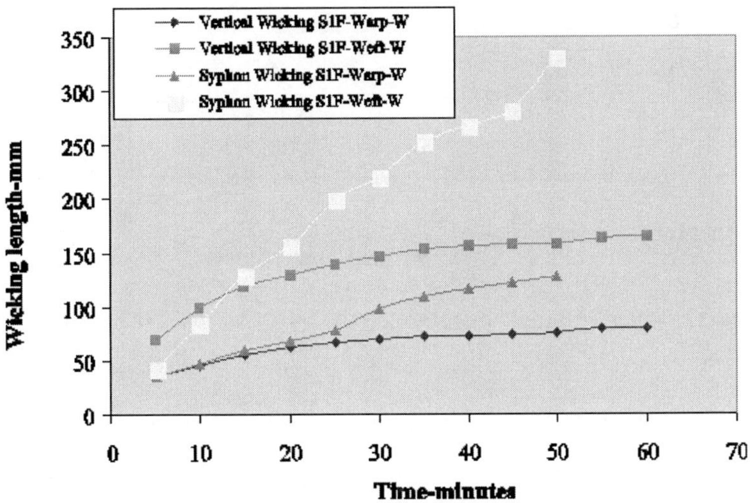

Figure 9. Actual Liquid Advance-Vertical Vs. Syphon Wicking Fabric S1F-Washed

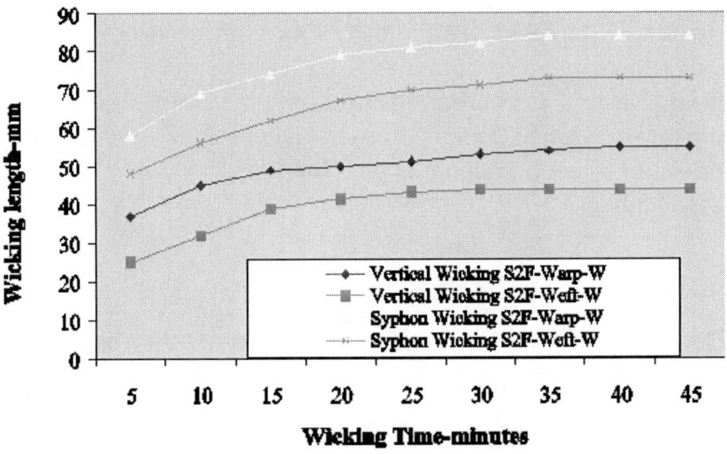

Figure 10. Vertical Vs. Syphon Wicking-Fabric S2F-Washed

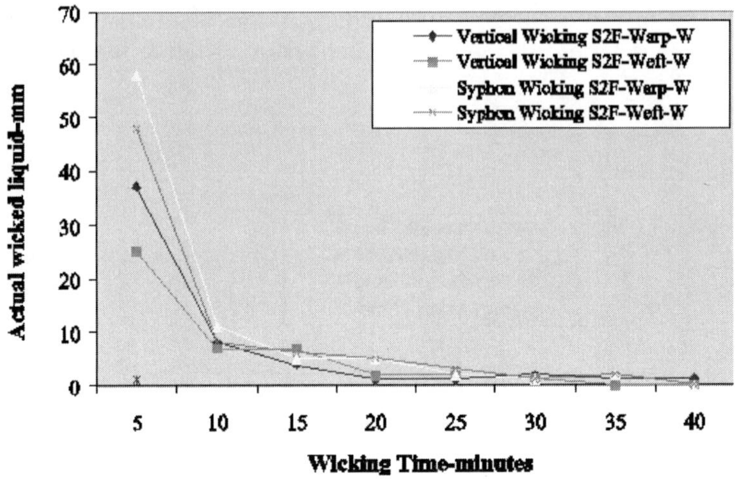

Figure 11. Actual Liquid Advance: Vertical Vs. Syphon Wicking Fabric S2F-Washed

6.1. Results and Discussion

6.1.1. Vertical Wicking Sample S1f- Washed Fabric: Warp And Weft Directions

Figures 17 to 20 show the graphical representation of the wicking rate of the washed and unwashed fabrics plotted from results in Table 2. Multiple comparison of the fabric wicking behaviour after a single wash in Table 5 show that there was a significant increase in the wicking rate of sample S1F in both the vertical and horizontal directions. In all cases, the weft direction wicking rate of washed fabrics remained higher than warp direction wicking regardless of the orientation of the fabrics as was the case with the unwashed fabric.

6.1.2. Vertical Wicking Sample S2f -Washed Fabric: Warp And Weft Directions

Table 2 and Figures 23 to 24 show the vertical wicking results of sample S2F. Results in Table 6 show that there was a significant increase in wicking in both directions after the fabrics were washed. Wicking in the warp direction was more rapid than in the weft direction and the difference gradually decreased with the lapse of time.

(a) Sample S1Y 44F34 Flat Fully Dull PA 6.6

(b) Sample S2Y 33F34 Flat Fully Dull PA 6.6

(c) Sample S3Y Air Textured Bright PA6.6

Figure 12. Nylon 6.6 Yarn Micrographs

6.1.3. Horizontal Wicking Of Washed Fabrics

Fabrics wicked in the horizontal direction (Table 2 and Figures 13 and 19) show a similar change in wicking trend as the fabrics wicked in the vertical direction. Figures 15 to 18 show that there was marked increase in the wicking rate of samples S1F and S2F after a single wash.

Results of multiple comparison of the actual liquid wicked within the 1st and 2nd quarters of an hourly test for fabric S2F exhibited a significant decrease in the liquid wicked in the horizontal direction compared to wicking in the vertical direction. This deviation from the general trend that horizontal wicking leads to a significant increase in wicking could not be explained.

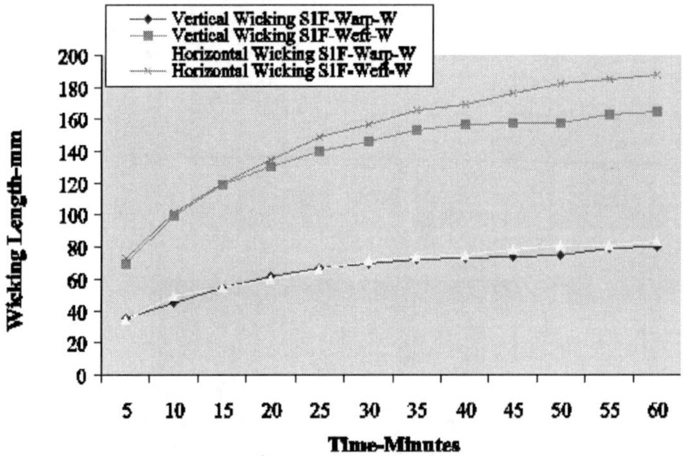

Figure 13. Wicking Tests Sample S1F- Washed Fabric

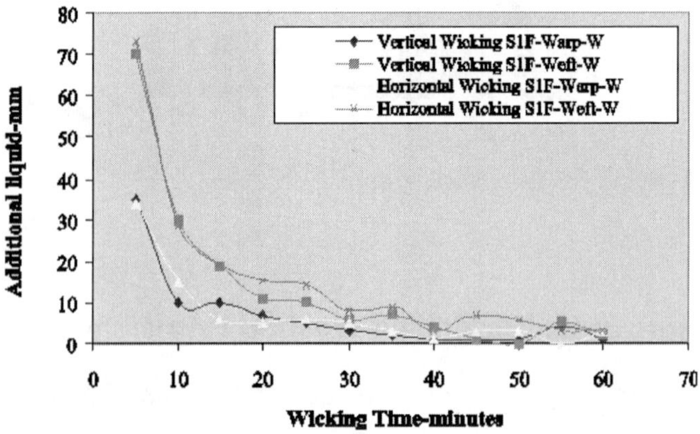

Figure 14. Actual Liquid Advance Sample S1F-Washed Fabric

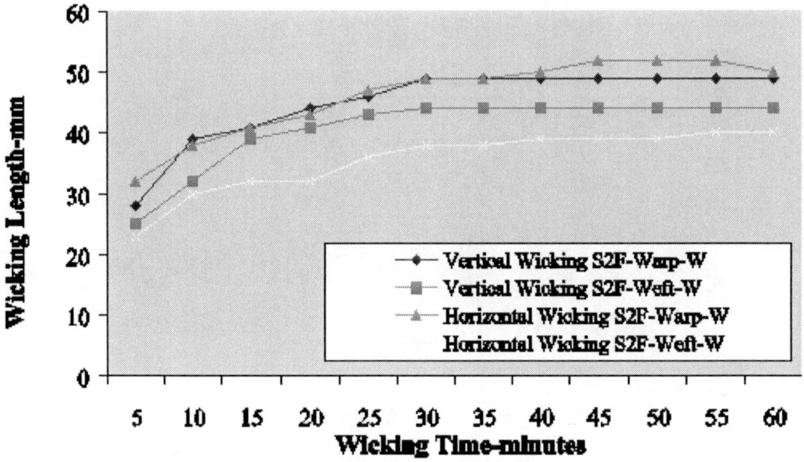

Figure 15. Wicking Tests Sample S2F-Washed Fabric

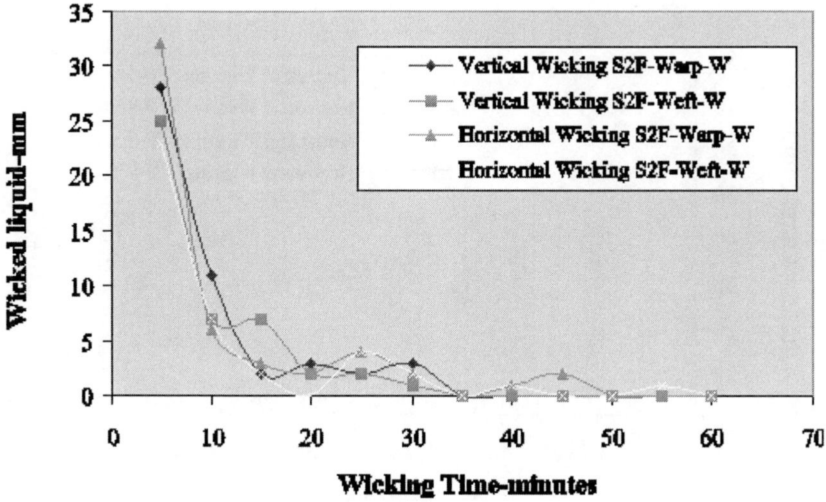

Figure 16. Actual Liquid Advance Sample S2F- Washed Fabric

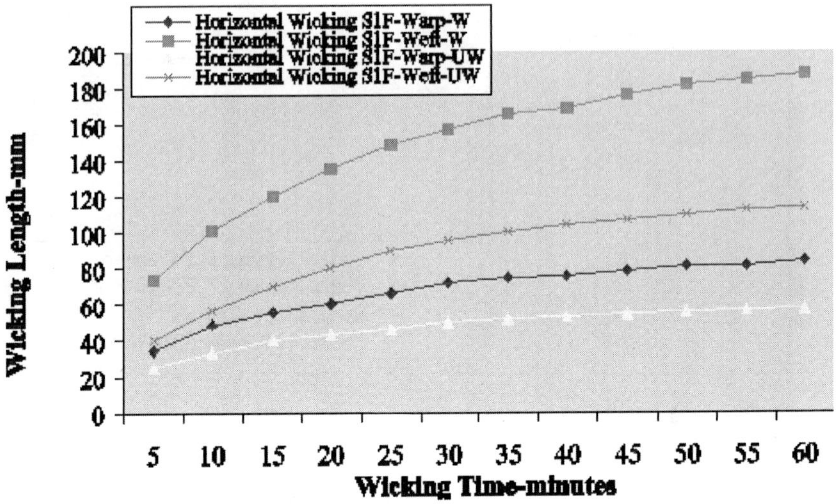

Figure 17. Wicking Tests Sample S1F-Washed Vs. Unwashed Fabrics

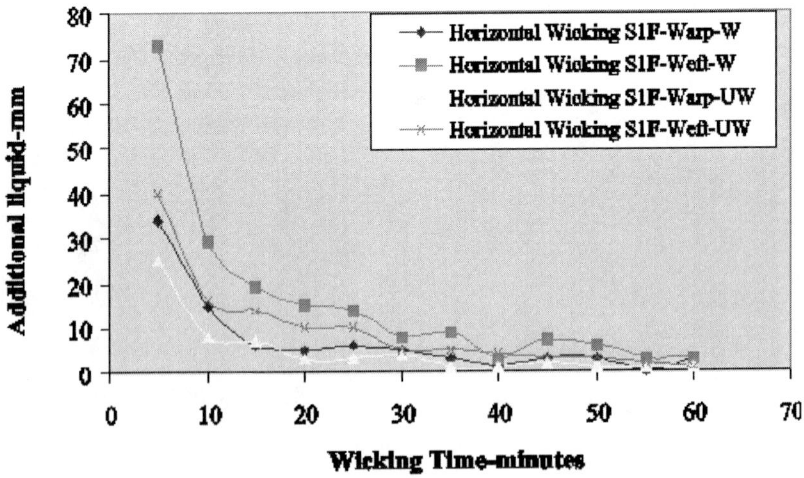

Figure 18. Actual Liquid Advance Sample S1F-Washed Vs. Unwashed Fabrics

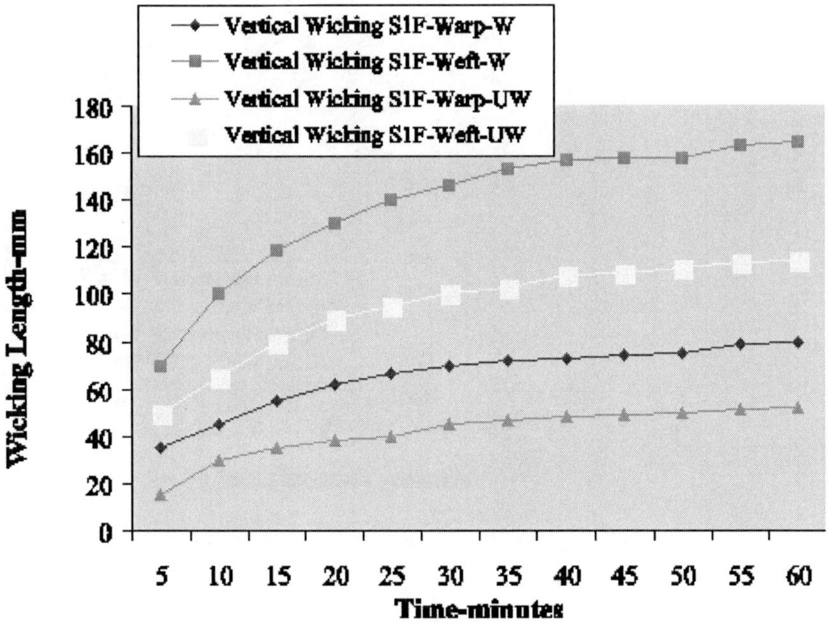

Figure 19. Wicking Test Sample S1F- Unwashed Vs Washed Fabrics

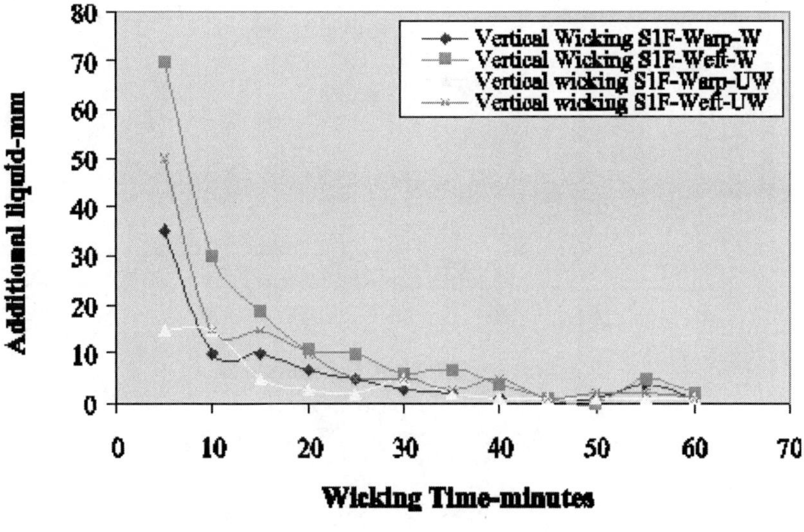

Figure 20. Actual Liquid Advance Sample S1F -Washed Vs Unwashed Fabrics

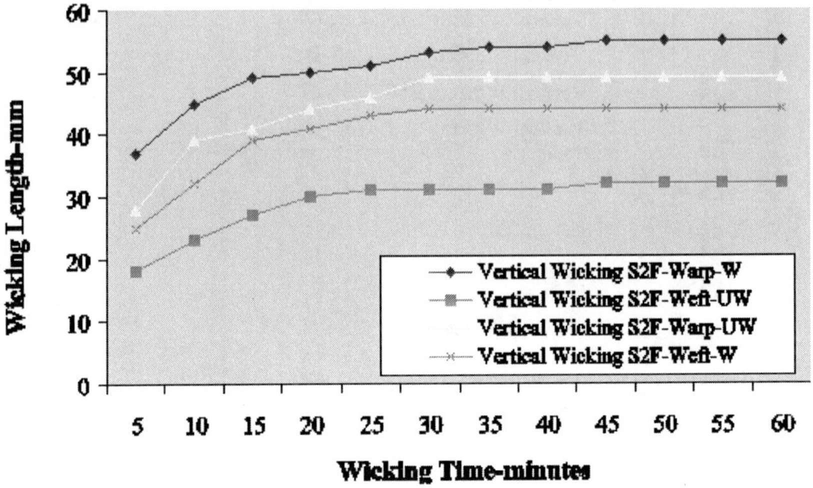

Figure 21. Wicking Tests Sample S2F-Washed Vs. Unwashed Fabrics

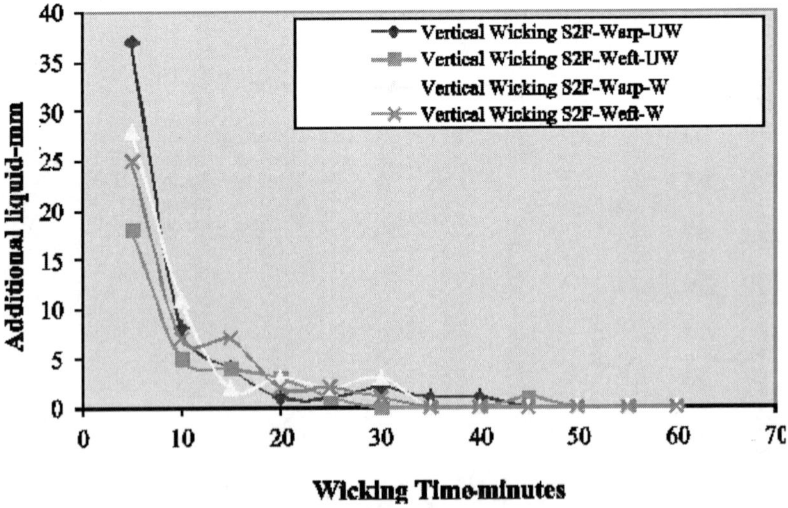

Figure 22. Actual Liquid Advance Sample S2F -Washed Vs Unwashed Fabrics

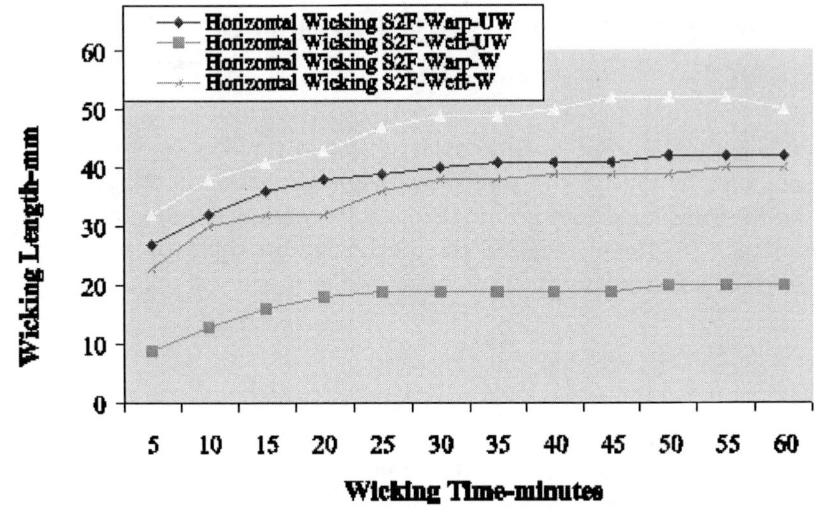

Figure 23. Wicking Tests Sample S2F-Washed Vs. Unwashed Fabrics

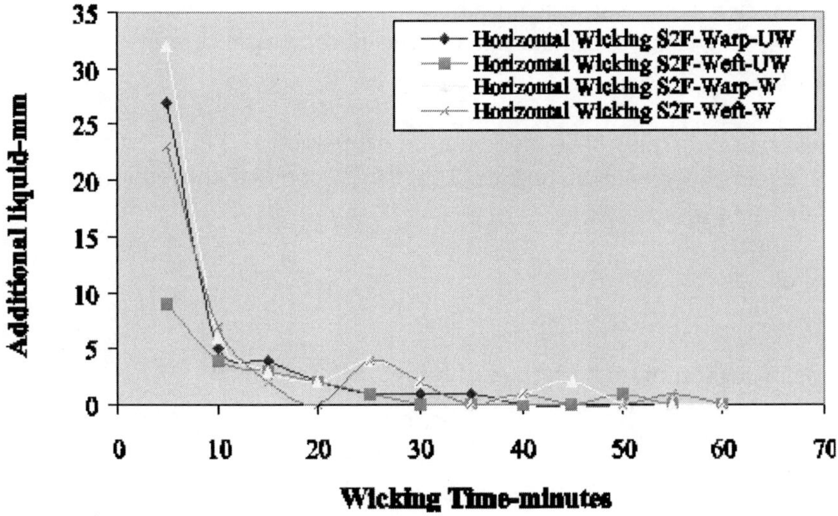

Figure 24. Actual Liquid Advance Sample S2F -Washed Vs Unwashed Fabrics

7. CONSISTENCY WITH WASHBURN'S EQUATION: FABRICS

The general laws that govern capillary flow in simple cylindrical tubes as expounded by Washburn's well-known equation shown in (1) is frequently used to study liquid transport in textile substrates as information obtained from such treatment is useful for the qualitative characterization of the process of liquid transport[13] in complex textile structures.

$$h = Ct^{\frac{1}{2}} \tag{1}$$

Where *h* is the distance travelled by a liquid in time *t* and C is proportional to the set of factors

$$\left(\frac{\gamma r \cos\theta}{\eta}\right)^{\frac{1}{2}} \tag{2}$$

Where γ = liquid surface tension, η = viscosity of the wicking liquid, θ = contact angle of the liquid against the fibre substance and *r* = capillary radius.

Several researchers have modified the expression as a basis for calculation of liquid movement in textiles. Laughlin[19] modified the equation into a general form

$$h = ct^{k} \tag{3}$$

Taking logarithms of both sides of this equation gives

$$\ln(h) = k\ln(t) + \ln c \tag{4}$$

This equation has the form of a straight line.

Plots of the logarithm of the height of rise *h* and the logarithm of the duration of time *t* in Figures 19 to 26 have a form of a straight line indicating that the wetting liquid follows diffusive capillary dynamics.[20] The tabulation of the *k* values of fabric S2F made from flat continuous filament yarns given in Table 8 ranged from 0.1487-0.2925 and for fabric S1F composed of continuous filament warp and textured filament weft yarns the range was from 0.3312-0.4427. In all the cases the time exponents*k* were less than Washburn's predicted time exponent of 0.5. which was attributed to the non-uniformity of the weft filament arrangement and the simultaneously occurrence of wetting, wicking, liquid dispersion and evaporation. Data points deviating from the trend line (Figures 25-32) mostly towards the end is an indication that with a significantly volatile liquid like water, evaporation from the wet surface of the fabric strip can compete with capillary process that advances the liquid.[12]

Table 8. Strip Wicking Test k-values

Sample	Description	Vertical wicking k-value	Horizontal wicking k-value
S1F-warp direction	Unwashed Washed	0.4427 0.3262	0.3255 0.3478
S1F-weft direction	Unwashed Washed	0.3312 0.3277	0.4217 0.3773
S2F-warp direction	Unwashed Washed	0.1487 0.2051	0.1725 0.1965
S2F-weft direction	Unwashed Washed	0.2179 0.2133	0.2925 0.2125

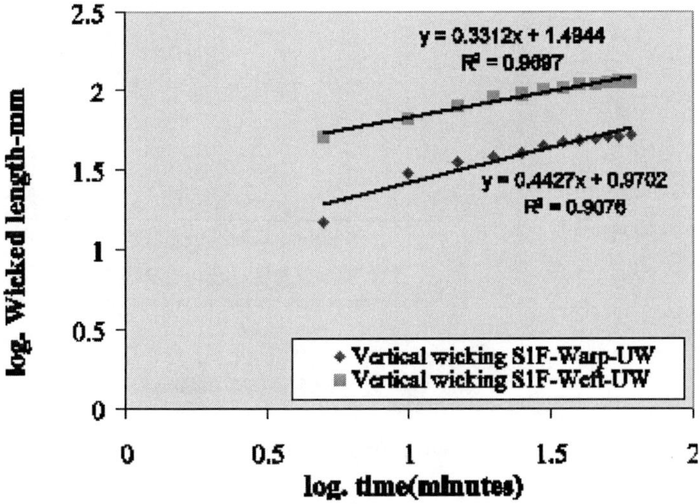

Figure 25. Vertical Wicking Sample S1F-Unwashed Fabrics

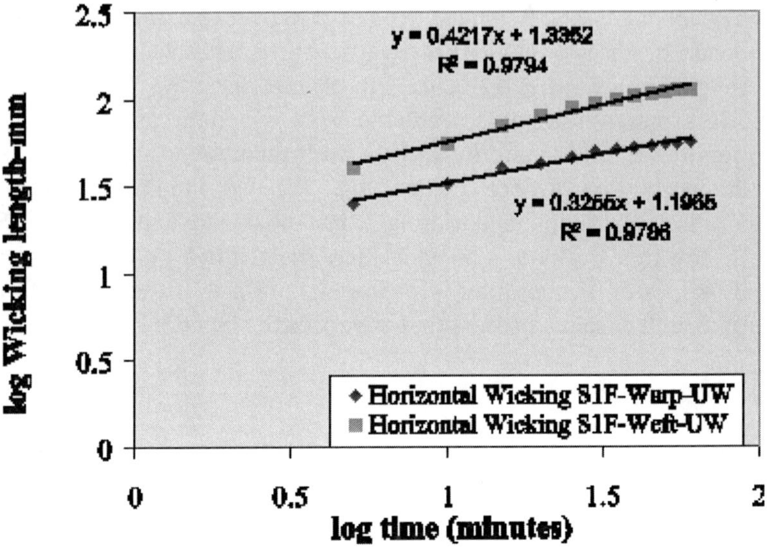

Figure 26. Horizontal Wicking of Sample S1F -Unwashed Fabrics

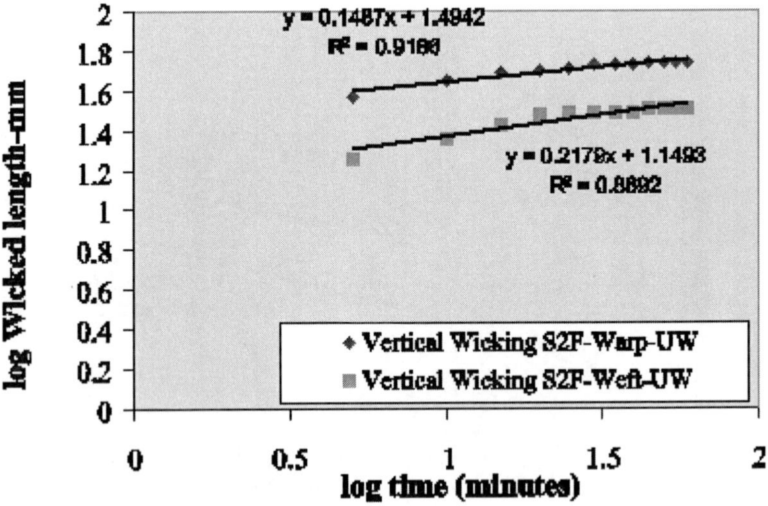

Figure 27. Vertical Wicking Samples S2F-Unwashed Fabrics

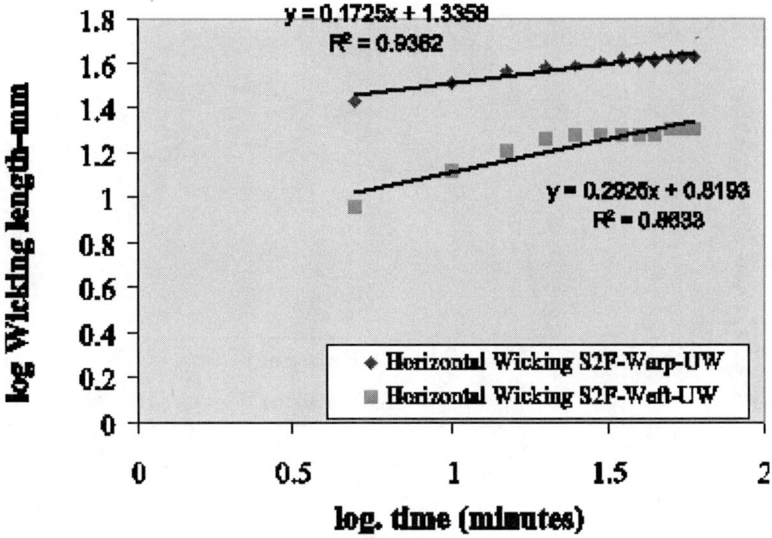

Figure 28. Horizontal Wicking Sample S2F-Unwashed Fabrics

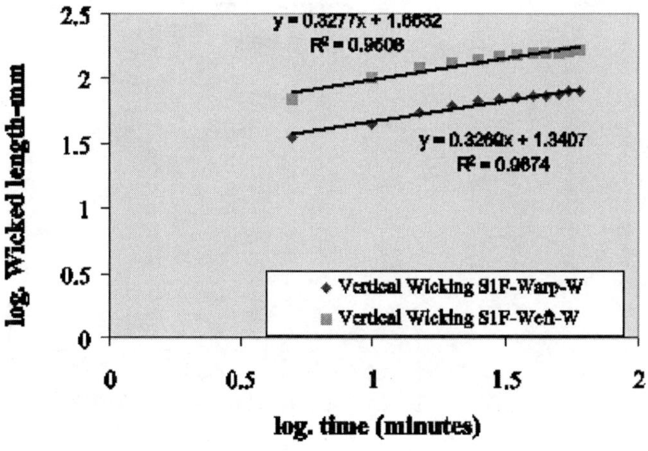

Figure 29. Vertical Wicking Sample S1F-Washed Fabrics

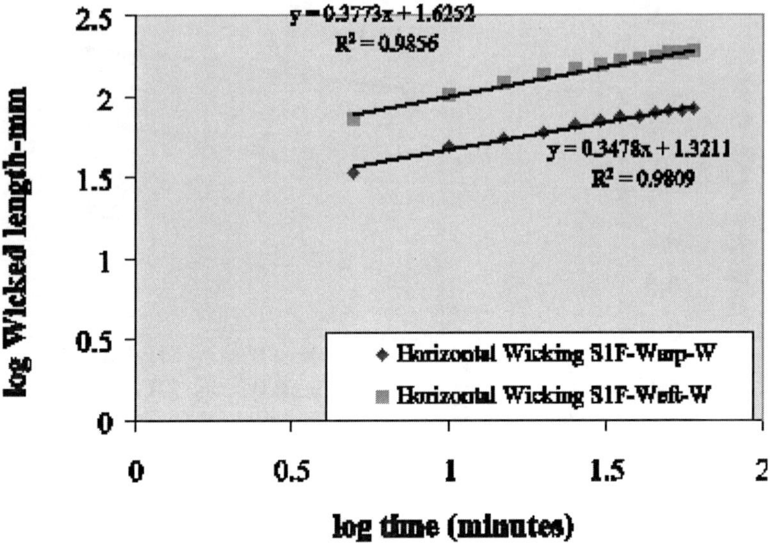

Figure 30. Horizontal Wicking Sample S1F-Washed Fabrics

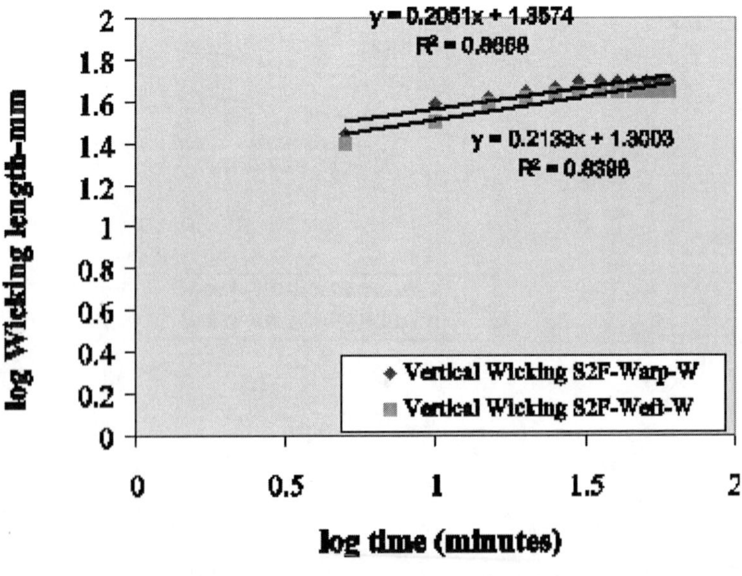

Figure 31. Vertical Wicking Sample S2F-Washed Fabrics

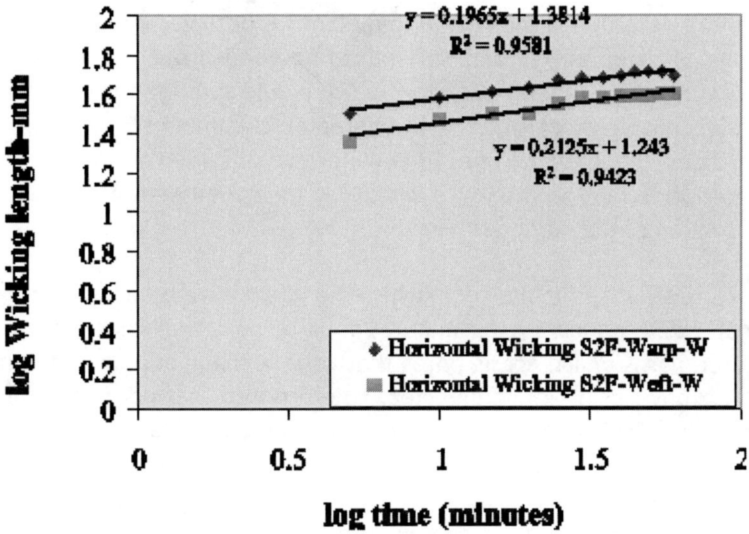

Figure 32. Horizontal Wicking Sample S2F-Washed Fabric

8. CONCLUSION

Miller and Tyomkin[21] state that when a porous material such as a fabric is placed in contact with a liquid, spontaneous uptake of liquid may occur. Law[9] observed that if the wicking distance is plotted against time, the graph is expected to have an initial rapid rate of change which decreases subsequently because water is first sucked into wider capillary channels by the action of surface tension. As the wicking process proceeds further, the total viscous resistance to the flow increases and the rate of flow decreases. In the case of the vertical strip test, the height and the mass of the water absorbed in the sample strip will gradually reach a quasi-equilibrium state when they are balanced by the hydrostatic head of water. In the case of the horizontal strip test, if the supply water is unlimited, the rate of penetration will gradually become constant.[9] In thick fabrics vertical wicking would continue with little effect of evaporation until a quasi-equilibrium state is reached when the wicking level in the fabric is balanced by gravity.[10]

In this work vertical and horizontal wicking of samples S1F and S2F did not continue indefinitely indicating that due to the combination of low fabric weight and thickness the maximum wicking height was not only influenced by gravity but also by evaporation. The rate of evaporation of liquid therefore determined the equilibrium point for both vertical and horizontal wicking of samples S1F and S2F indicating good properties required for eliminating perspiration discomfort which would cause fabric wetness with resulting problems of freezing in winter or clamminess[22] in summer. In most cases, the leading front of the water rise observed at the end of each test period felt dry to the touch which can be attributed to the rapid liquid evaporation of the fabrics.

In textured yarns, the manner in which the liquid is transported through the fabric is determined by the minute loops or coils that characterize air – textured yarns which act as pores that vary in shape and distribution and may or may not be interconnected. Hsieh[6] noted that pore variation and distribution leads to preferential liquid movement towards smaller pores, resulting in partial draining of previously filled pores in the fibrous structure. In all cases studied in this work, tests showed that there is a good linear relationship between the logarithm of the wicked liquid (ll) and the logarithm of the wicking time (*t*) indicating that the wetting liquid follows diffusive capillary dynamics[20] even though for sample S1F in most cases the exponential values were high compared to sample S2F due to evaporation from the parallel packed filaments of the yarn structures.

The high k values of fabrics containing textured weft yarns indicate the characteristics of a non-homogenous capillary system where wicking is a discontinuous process due to the irregular capillary spaces of varying dimensions.[11] Rapid wicking is retarded by the 'absorber' textured weft yarns which are more bulky and act as temporary liquid reservoirs as all the voids are filled up. On the other hand, the inter-filament wicking rate is increased once the liquid is transferred to the flat 'runner'continuous filament warp yarn due to capillary sorption[11] resulting in spiked wicking behaviour observed.

Wicking is also affected by fabric construction. Fabric sample S2F wicked more rapid in the warp than in the weft direction due to the high density of ends in the fabric. If the filament packing in the yarn is assumed to be an idealized or closely packed assembly [23] there will be more capillaries in the warp than in the weft direction due to the distribution in the number of ends and picks.

Outdoor active wear such as jackets are infrequently washed and research24 results have shown that a standard 5 washes of vests used for mountaineering resulted in a significant increase in their wicking performance. Even though a spin finish was applied to fabrics S1F and S2F during finishing to give surface properties which can allow liquid flow, the durability of the spin finish to washing was insignificant since laundering of fabrics resulted in a significant increase in their wicking performance. Washing therefore did not lead to the collapse of the capillary system of the fabric but results in the re-arrangement of the capillaries between filaments due to the washing liquid movements and the relaxation of the textile structure during drying.[24]

REFERENCES

1. Barnes J.C and Holcombe B.V., Textile Res. J., 6612 777-786, 1996
2. N. J. Brownless, S. C. Anand, D. A. Holmes, T. Rowe, TextJ.Inst 87 Part 1, 1 1 172 182.
3. N. J. Brownless, S. C. Anand, D. A. Holmes, T. Rowe, TextileAsia.August 1996 77 80.
4. K. Slater, ComfortProperties.ofTextiles.TextileProgress.Volume.Number 4 4191 1 91 , Textile Institute 1977.
5. H. N. Yoon, A. Buckley, Res. Text, 54 289 298.
6. Hsieh. You-Lo, Res. J. Text, 655 299-307, 1995
7. N. J. Brownless, Anand. D. A. S.c, Holmes, T. Rowe, Sports. World, Volume. Activewear, 22 2 36-38, 1996
8. A. B. Nyoni, D. Brook, J. Text, Inst, 97 2 2006, 119-128.
9. Law Y.M.M., Ph.D Thesis, University of Leeds, 1988
10. Q. Zhuang, D. Ph, University. Thesis, of, Leeds, 2001
11. E. Kissa, Res. Text, 66 66 (10), 660-668, 1998
12. Miller B., International Nonwovens Journal, Volume 91 1 Spring 2000.
13. K. Pronoy, Chatterjee and Hien V. Nguyen., Mechanism of Liquid Flow and Structure Property Relationships., Absorbency, Chapter II., Edited

by Pronoy K. Chatterjee, Elsevier Scientific Publishers; Amsterdam; New York, NY: 1985

14. P. R. Harnett, P. N. Mehta, Res. Tex, 54, 471-478, 1984

15. A. B. Nyoni, 2003 PhD Thesis, University of Leeds.

16. Hepburn C.D., PhD. Thesis, University of Leeds 1998

17. A. Leijala, J. Hautojarvi, Res. Text, J.,68(3, 193202 1998.

18. G. T. Blyth, D. Ph, University. Thesis, of, Leeds, 1984

19. R. D. Laughlin, J. E. Davies, Res. J. Text, 31 904 910 1961

20. Perwuelz. Anne, Casetta. Mthilde, Caze. Claude, Testing. Polymer, 2. Volume,. Issue, 553 561.

21. B. Miller, I. Tyomkin, Res. J. Text, 5. Volume, 706 712 Nov. 1984

22. W. H. Rees, Month. Text, 59 61 August 1969

23. Hearle J.W.S., Grosberg P., and Backer S., Structural Mechanics of Fibres, Yarns, and Fabrics. 1 1969, John Wiley; New York, NY, USA.

24. A.B Nyoni and D.Brook, Textile Research Journal,Vol.80(8), 2010 720-725.

Chapter 6

Analysis of stress–strain of architect woven fabric strength under biaxial extensions

Magdi El-Messiry, Shaimaa Youssef

Textile Engineering Department, Faculty of Engineering,
Alexandria University, Egypt

ABSTRACT

Architectural fabric, or tensile fabric, structural fabric has become popular over the last decades. The structural fabric is used for shading public areas such as airport, boat docks, sport areas, public bus stops, and car parking. The most important advantages of the architectural fabric structure are low weight and more economical designs. In this research, three different fabric materials were studied to investigate the strain and stress in the fabric during time of using in the shading. The present work focuses on 2D biaxial orthogonal woven structure, among which plain weave is the distinct category. 2D means that the woven fabric features only in-plane reinforcing properties. Biaxial means that yarns are aligned along two directions: the warp and weft directions. Orthogonal means that the warp and weft yarns are laid down at 90° with respect to each other. The present work is aimed at developing a theoretical approach for prediction of fabric strain–stress relation based on the results achieved by experimental studies.

KEYWORDS

Architectural; Biaxial extensions; Orthogonal

1. INTRODUCTION

Architectural fabric structure is a new architectural form which replaced traditional structures because of its wide field of usages and its famous advantages. The architectural fabric structure consists of two main items. The first one is a steel frame to form the required shape of the design, and the second item is a tensioned fabric [1]. Some architectural fabrics are made of synthetic materials as woven glass fabric with a Polytetrafluor Ethylene (PTFE) coating or woven polyester fabric with polyvinylchloride (PVC) coating. The widely fabric material used is a polyester fabric with (PVC) coating which is utilized for a variety of reasons including its low cost and resistance to damage during fabrication and erection, but there are disadvantages of using woven polyester fabric with polyvinylchloride (PVC) coating as high levels of creep can necessitate re-tensioning and lower resistance to dirt [2]. Advantages of architectural fabric structures are more economical in comparison with standard building design [1], have low weight of materials than normal building design, especially in open spaces to be covered. Architectural fabric structure is flexible to be designed with various shapes to give beauty appearance in shading. Distribution of multi directional force in a woven fabric was studied by several investigators either theoretical or to develop a measuring instrumentation to measure the stress strain curves [2], [3], [4], [5] and [6]. In some applications of the architectural fabric, it may extend in multi directions because it is tensioned under a central concentrated force and other forces at the edges of the fabric.

Wasiak et al. [3] introduced a new method of measuring multi directional force distribution in woven fabric. The measuring system consists of mechanical part and electronic part. The mechanical part consists of grips positioned along circumference of the fabric sample and another part for loading the tested sample by a central force. The electronic part includes force transducers to measure the changing in the force at the circumference of the tested sample [3]. In such case the different forces applied on the fabric result not only in a tension in one part, but also a compression on some other areas of the fabric. The shear properties of the fabric will play a certain role in defining the value of the strain in such cases. Several researchers [4], [5] and [6] studied shear property with different woven fabric structures, especially plain woven fabric, and found that the fabric shear modulus is less in plain woven fabric structure than in the other woven fabric structures. The main advantages offered by textile reinforcements are: lower manufacturing costs, easier handling and processing, better forming in complicated geometrical shapes of the shaded part, and the possibility of applying a variety of processing techniques.

2D architecture fabric in 3D structures results in complex strain–stress field in the architect fabric material, leading to new failure mechanisms and failure modes not seen in traditional unidirectional laminates. It becomes known that a fabric is an extremely complex structure for mechanistic analysis. There exist several structural levels from fibers to yarns and eventually to the fabric. Each level has its own geometrical and mechanical variables which control or influence to varying degrees the fabric behavior. The objective of this work is to study the relation of the strain–stress contours distribution in a tensioned fabric under a central force and the method to reduce it.

2. MATERIAL AND METHODS

2.1. Material

To measure the strain distribution, fabric samples (45 cm × 45 cm) were prepared and marked with a grid (5 cm × 5 cm) for measuring the strain in the fabric after being tied on the apparatus. Table 1 shows fabric specifications of different samples used in this research.

2.1.1. Apparatus

An apparatus for testing architectural fabric in different space structure forms is based on the design given by Pan [10]. Fig. 1 shows architectural fabric structure forced from the center using rod of 5 cm diameter, so the central force (*F*) can be applied on the sample center by various values through the hydraulic arm.

Table 1. Fabric specifications of different samples.

Sample no.	Fabric design	Fabric material	Fabric weight (gm/m^2)	Fabric thickness (mm)	Ne_1	Ne_2	Ends (in.)	Picks (in.)	Fabric tightness
I	plain	100% cotton	316	1	10/2	7	40	28	0.57
II	plain (duck)	100% cotton	590	1.75	3	6	45	40	0.79
III	plain	100% glass fiber	106	1.5	0.6	0.6	15	15	0.30

Figure 1. Architectural fabric structure testing set-up.

The sample, as mentioned previously, was tied on the apparatus with the required configuration. The change in the grid length lines was used to calculate the strain contour in each part of the fabric under the applied central load.

3. RESULTS AND DISCUSSION

3.1. Strain contour on the fabric under central load

Recently, woven fabrics have been applied as the fiber performs for fiber-reinforced composites. Consequently, it becomes increasingly desirable to develop a more advanced mechanistic theory governing the fabric responses under various loading situations so as to provide a powerful design tool for fabric structures in heavy load-carrying applications. The different fabric samples were tested and the strain contours in each case were measured. Fig. 2 illustrates the distribution of strain contour in weft direction for fabric sample (II) under central force (*F*).

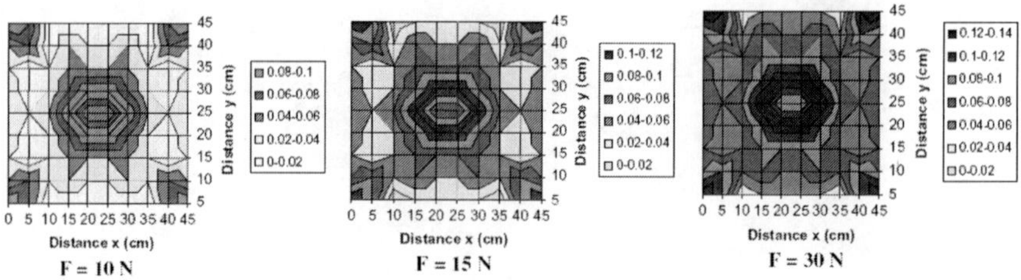

Figure 2. Contour curves of strain in weft direction for fabric.

So far, attempts to investigate the mechanical behavior of woven fabrics have more or less followed three paths. The well-known Peirce geometrical model [7] has been considered to be the first such undertaking. By assuming a woven fabric as a highly idealized geometrical object, Peirce was able to describe the deformational behavior of the fabric under external loading. Painter [8] and Love [9] further refined this technique in their analyses of fabric properties. The interlacing points are the major locations where interactions between yarns in the two bundle systems take place, through which the yarns form an interlocked structure. Pan [10] indicates that without the interactions occurring at the interlacing points, a woven fabric would be equivalent to a system of two sheets, each made of parallel but isolated yarns; the resultant properties would be entirely different from those of a practical fabric. As stated above, the yarn to yarn interactions in a woven fabric mainly take place at the yarn interlacing points through yarn to yarn contact. However, the contact area at an interlacing point between a warp and a weft yarn is only partial, as shown in Fig. 3 of an idealized illustration of a plain weave [10]. If an external tension σ_y is applied to the fabric, force equilibrium can be established at a differential portion d_x as in the following equation:

$$\frac{L_y}{2}\tau_y n_y d_x = d\sigma_y \tag{1}$$

where σ_y is the external tension, τ_y the shear resistance, n_y the number of yarns in ydirection, and $L_y/2$ is the actual contact length between yarns, for one contact point which depends on the dimensions of the yarn in both directions as given by the following equation:

$$L_y = 1.51\pi(t_y + w_y) - \sqrt{(t_y w_y)} \tag{2}$$

where t_y is the shorter axes of the yarn cross section, and w_y is the longer axes of the yarn cross section.

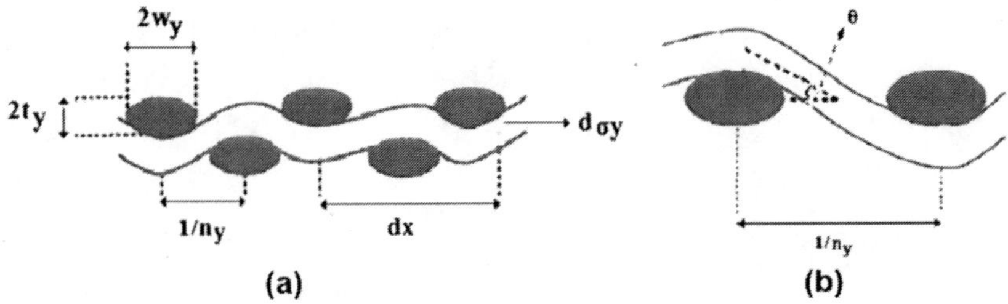

Figure 3. Forces on the plain weave under tension [10].

Pan [10] shows that the components of the shear resistance τ_r at the yam contact area, even with a fabric made of yarns whose weight is so light as to be negligible, a certain force is still required to pull a yarn out of the fabric by overcoming the resistances between the yarn contact points. Yet the magnitude of this pullout force will increase significantly when certain pressure is applied at the contact points. In other words, there exist two components of the shear resistance τ_y as in Eq. (3) which resist any attempt of relative yarn movement at the contact area. Of the two components, one is pressure-independent adhesive force termed τ_{y2}, and the other is pressure-related frictional forceτ_{y1} governed by the friction law, as explained above, that is;

$$\tau_y = \tau_{y1} + \tau_{y2} \tag{3}$$

where τ_y is the shear resistance, τ_{y1} the pressure-related frictional force governed by the friction law, and τ_{y2} is the pressure-independent gripping force termed.

The pressure can either be applied directly to the fabric surface, or be generated from tensile load exerted uni-axially or biaxially to cause a tightening effect to the fabric so as to increase the pressure at the yarn contact points. Assume the fabric is under a bi-axial loading case where the load σ_L is applied in the warp direction and the load σ_T in the weft direction as seen in Fig. 4a, these external loads will create a pressure (*P*) at each contact point of a cross-section of the fabric, as shown in Fig. 4b and c.

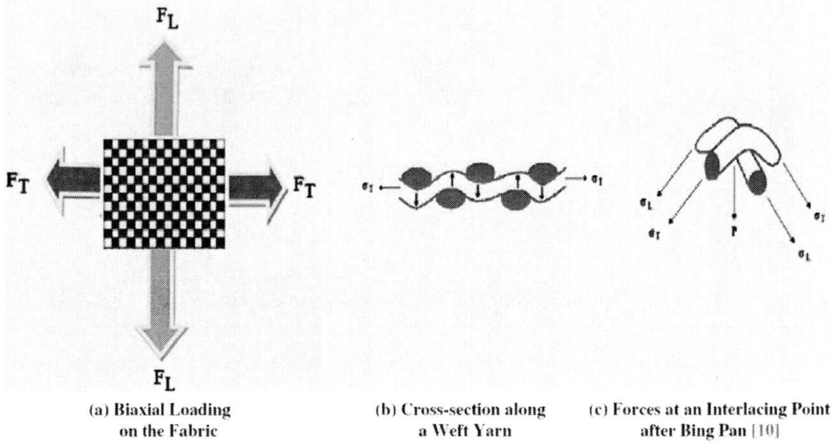

Figure 4. Various forces on a plain weave fabric.

In the case under the study, strain contour on the fabric is found to increase from the edge to the center of the fabric with highest value at the apex of the fabric shape.

With the increase of the center force (*F*) the apex angel will decrease, thus the force in the direction of the supports increases too.

$$R = (F/4)/\cos(\Phi/2) \tag{4}$$

where *R* is the force in the direction of the supports, *F* the central force, and Φ is the apex angle.

Fig. 5 shows that, as the central force increases, the fabric strain will increase, mostly in ring shapes, because the steel center rod which pushes the fabric from the center to up has a circular shape. The most fabric strain happened at the fabric edges and in the center. It is due to the tension from tying at each edge of the fabric.

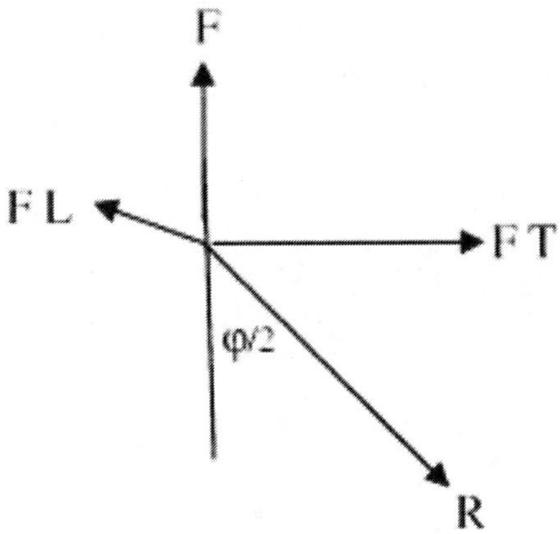

Figure 5. Analysis of the forces on fabric.

From the above analysis, it can be concluded that the value of the stress in the warp and weft directions will depend on the value of the forces applied on the fabric and the shear force, as well as the coefficient of friction between the weft and warp yarns. The yarns in each direction will take a curved shape depending on the resultant force on each yarn. The curvature of each warp or weft yarn will determine the value of strain. Fig. 6 indicates that the value of strain reaches maximum at the line of application of force *R*, and then decreases in the direction to the middle of the base line of fabric triangle on each side. The value of the strain decreases in the direction of the fabric edge and in some cases it will be of negative values. The maximum value increases as the central force increased, but not in proportional values. This is due to the fact that the fabric will redistribute the load reducing the maximum strain value. The minimum strain value may be negative at low values of central force.

Figure 6, Figure 7 and Figure 8 show the change of strain distribution in weft and warp directions for the different fabric samples.

3.2. Stress analysis in the architectural fabrics under central force

It is obvious from Fig. 9 that stress contour has the same behavior as strain contours given in Fig. 2. However, the increase of stress is not proportional to the increase of the value of central force applied, again because the redistribution of the strain all over the area of the fabric pre increases the central force (*F*).

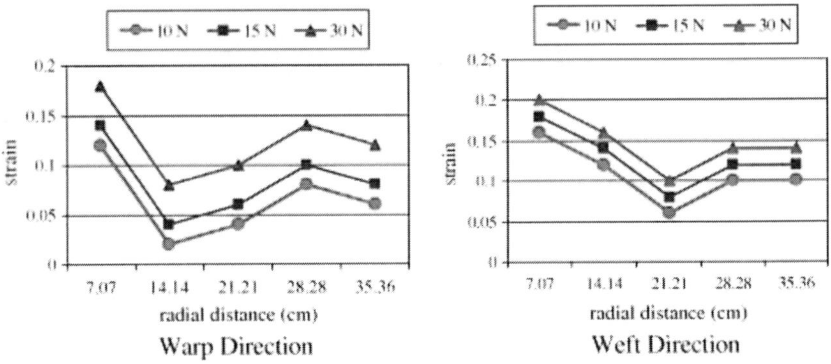

Figure 6. Effect of different central forces on fabric strain (I).

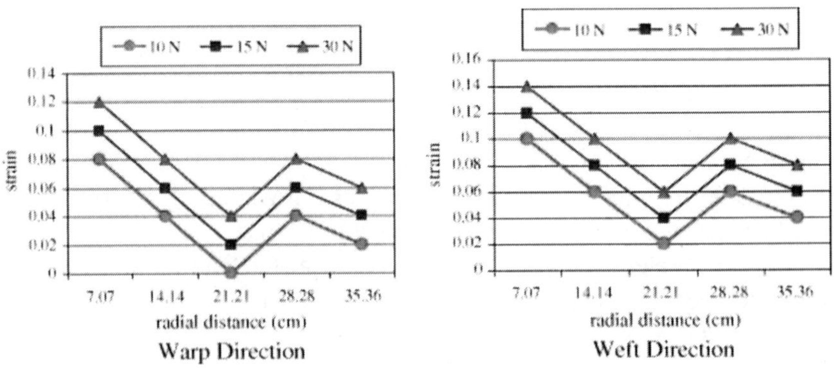

Figure 7. Effect of different central forces on fabric strain (II).

The value of stress will depend on the strain in the warp or weft direction and fabric young's modulus, thus the sample (I) shows the higher stress value in warp direction than that in weft direction. This is because the yarn count and ends per inch in these directions are not the same.

Fig. 10 shows the fabric stress in warp and weft directions happened in a quarter part of the three different fabric samples (I, II, and III) with different configurations.

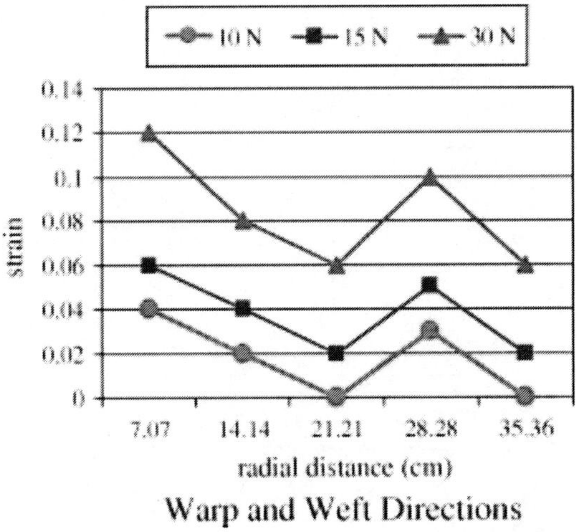

Figure 8. Effect of different central forces on fabric strain (III).

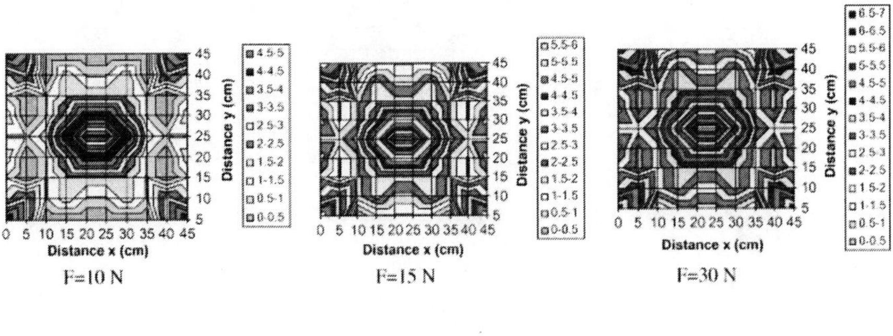

Figure 9. Contour curves of stress in weft direction for fabric (II).

In addition, Fig. 10 points out that the values of fabric stress of weft yarns are less than of the warp yarns. It is because of the difference in counts where weft yarns are finer than warp yarns and the weft thread density is less than the warp thread density in the both fabrics (samples I and II).

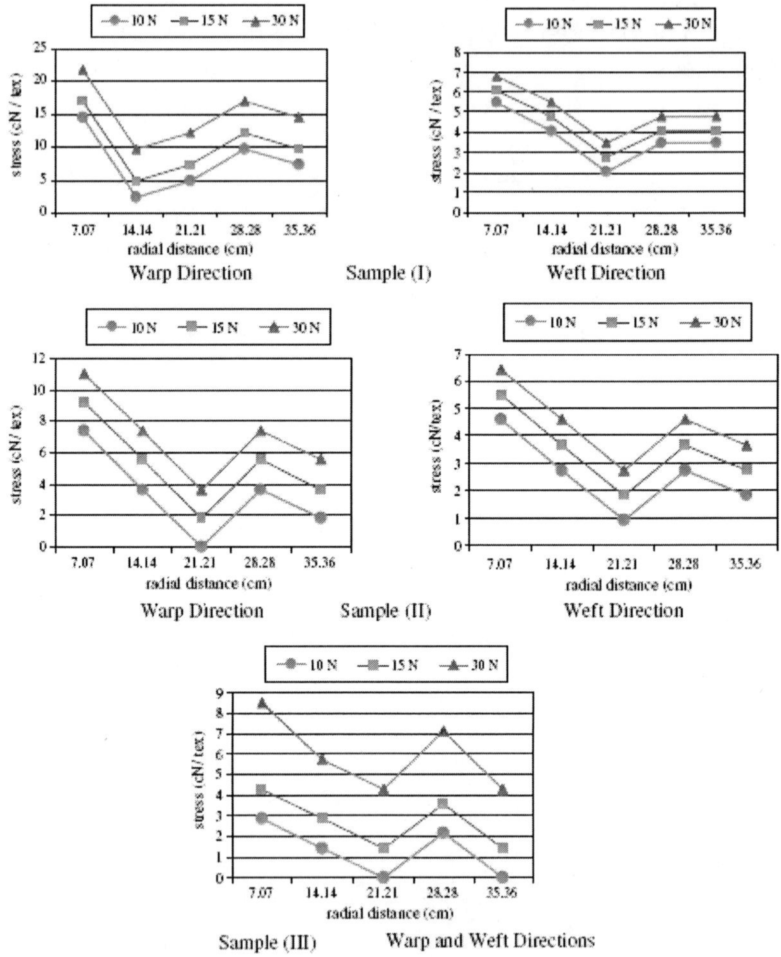

Figure 10. Fabric stress in warp and weft directions.

3.3. Stress reduction in the fabric

The above analysis the fabric samples designate that when they are loaded as one piece to form 3D structure the stress distribution varied from the maximum value at the line of the application of the force (*R*) to the minimum value at the middle of the fabric base line. In order to reduce the value of stress in the architectural fabrics under a central load, the fabric sample was cut into four parts in triangle shapes and sewed together using four tapes and attached on the apparatus with central force (30 N), as shown in Fig. 11.

Figure 11. Strengthened edge sample.

This strengthened edge sample was compared with another similar fabric samples without sewing taps attached.

Fig. 12 illustrates that the method using strengthened edge sample leads to lowering fabric strain and stress compared to the traditional method. The using of the tapes at the fabric edge will reduce the fabric strain and stress at this line.

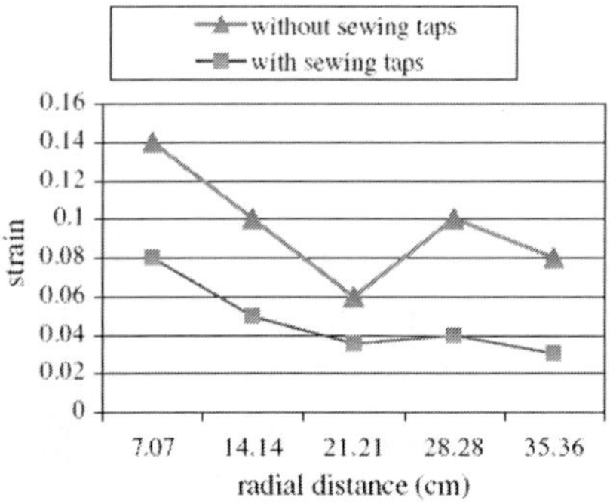

(a) Fabric Strain in Weft Direction

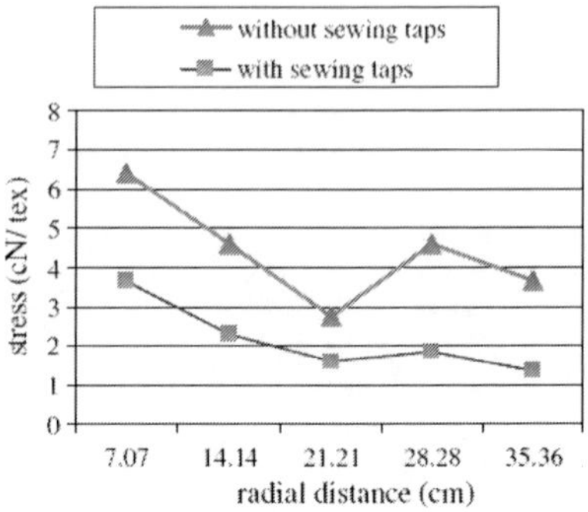

(b) Fabric Stress in Weft Direction

Figure 12. Effect of using strengthened edge direction for fabric (II).

4. CONCLUSION

This study designates that the stress and strain distribution in the architectural fabrics depends upon the type of fabric, its structure and mechanical properties. It is recommended to form the final 3D shape of the structure by tailoring the fabric into the parts and sewing them together with the seams with additional strengthened taps to reduce the maximum stress value in the fabric.

REFERENCES

1. http://www.tricoshade.com/fabricarchitecture.htm.

2. B.N. Bridgenes, P.D. Gosling, M.J.S. Birchall, Tensile fabric structures: concepts, practice & developments, Te Structural Engineer 20 (2004) 1–27.

3. I.F. Wasiak, M. Snycerski, Z. Stempien, H. Suszek, Measuring method of multidirectional force distribution in a woven fabric, Fibers and Textiles in Eastern Europe Journal 2 (46) (2004) 48– 51).

4. G.A.V. Leaf, A. Sheta, The initial shear modulus of plain-woven fabrics, Journal of Textile Institute 75 (3) (1984) 157–163.

5. M.A. El-Messiry, A. Sheta, Measurement of fabric shearability by twisting method, The Indian Textile Journal (1987) 122– 129.

6. A.C. Long, Design and manufacture of textile composites, Textile Institute, Woodhead Publishing Limited, 2005, p. 129– 31.

7. F.T. Peirce, The geometry of cloth structure, Journal of Textile Institute 28 (1937) T45.

8. E.V. Painter, Mechanics of elastic performance of textile materials: Part VIII: Graphical analysis of fabric geometry, Textile Research Journal 22 (1952) 153.

9. L. Love, Graphical relationships in cloth geometry for plain, twill and sateen weaves, Textile Research Journal 24 (1954) 1073.

10. Ning Pan, Analysis of woven fabric strengths: prediction of fabric strength under uniaxial and biaxial extensions, Composites Science and Technology Journal 56 (1996) 311–327.

Chapter 1

Functional Design of the Woven Filters

Lucica Cioară[1] and Ioan Cioară

[1] Technical University "Gheorghe Asachi"Iassy/Faculty of Textiles and Leather Engineering, România

1. INTRODUCTION

The filtration process implies the physical separation of one or more components of a fluid that passes through or over a barrier which is permeable to only one or some of the fluid components. Therefore the fundamental element of the filtration process is the barrier which is permeable to only a part of the suspension or the solution applied to filtration. This barrier is named filter medium and the mechanical structure used to support it is named filter. The statement the heart of any filter is the filter medium is fully justified. The most ingenious filter is useless if does not have an adequate filter medium. A specific shape of a filter can use a wide variety of filter mediums to do the same or different separation.

Function of their purpose filtration processes are used to separate solid – gas, solid – liquid, liquid – liquid or solid – solid mixtures. Solid – gas separation domain is represented mainly by air filters also including gas processing. Solid – liquid separation is the usual area of mechanical filters from which a relevant part is the inertial separators. Liquid – liquid and solid – solid separations are complex and specialized areas of filters or separators typology.

Industrial installations frequently use as filtration media technical textiles obtained by weaving i.e. woven filters (Adanur, 1995; Harracks & Anand, 2000). To respond to imposed exigencies by the use in industrial installations, the fabrics utilized as filter media must comply with a wide range of demands which are for the most part determined by the fabrics own structural characteristics and partially by the fabric finishing methods (Marchiş et al.,1991; Cioară et al., 1991; L.Cioară&I.Cioară, 2001). Among these requirements the following are mentioned:

- high filtration capacity, high degree of filtered elements purification and minimum hydraulic resistance;
- good mechanical resistance and stability to chemical, thermal, corrosive and biological agents;
- a high degree of the filtering surface smoothness allowing an easy and total residue separation and filter regeneration;
- a firm and homogenous structure allowing a high filtering process fineness and quality throughout the service life of the filtering element.

The fields having a vital requirement for woven textile media filtration are in a continuous diversification and specialization. It is widely known that technical textiles represent viable alternatives for all economic and social life sectors. In this context the woven textiles filter media, a significant representative of technical textiles, find their application in various fields from automobiles to space industries, in construction, in agriculture and environment protection (Adanur, 1995;Harracks&Anand, 2000).

Filtration processes are accomplished following two basic principles: depth filtration and surface filtration (Medar&Ionescu, 1986). Both forms of the filtering process imply the simultaneous occurrence, in different ratio of two physical phenomena:

- direct particles retention – do to this process filter medium stand for a mechanical barrier for particles bigger than the restriction;
- adsorption – particles retention by electrostatic forces or molecular attraction of the filter medium.

Principle of depth filtration and the specific mechanisms through which the particles are retained are shown in Figure 1.a. Direct interceptions take place when particles of a certain size are passing through larger pores and are trapped in the filter medium structure when meet smaller pores. Inertial forces cause particles to hit filter medium fibers the particles retention being obtained due to their penetration into the body of fibers or due to the fiber deformation.

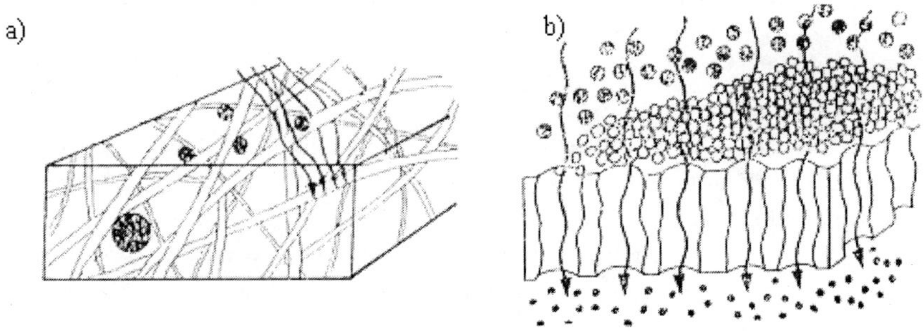

Figure 1. Filtrations mechanisms

Inertial impaction is predominant when high fluid velocity or very dense filter medium is present. This type of filtration mechanism is most predominant when high gas velocity and/or dense packing of the filter media is present. Inertial impaction occurs also when an abrupt change in streamline take place. In this case the particle, due to its inertia, will continue along its original path and could be retained by the filter medium

Adsorption phenomena determine the attraction of small size particles by the filter medium fibers. The adsorption is favored by particles Brownian movement of the particles during the filtration process. Textile filter media that work by the depth filtration principle are: nonwoven fibrous layers, simple textiles made of spun or filamentary yarn, pile or felted, composite fabrics made as semi double, double or multiple layers structures.

Surface filtering implies that particles larger than the pore size are retained on the filter medium surface (Figure 1.b) Due to the adsorption forces particles smaller than pore size can be retained along the pore wall, reducing its transverse dimension causing blocked pores and filter medium clogging as a result. In the first phase of the clogging nominal fineness of filtration is reduced, the pressure difference increases and a combination of surface filtration with a pseudo-depth filtration take place (occurs).

Later on, as the degree of clogging increases, fluid flow through the filter medium is significantly reduced. Textile filter media which operate by surface filtration are monofilament yarns woven textiles.

The comparative analysis of the woven filter media working according to these two principles highlights their advantages and disadvantages (Table 1). In all cases the filter media is considered within the conventional filtering range ensuring the separation of particles over 1 μm in size (Rouette, 2001).

Filtration fineness is influenced by filter medium structure. Pore size distribution is Gaussian for filter media that operates on depth filtering principle and covers a narrower range around the mean value for filter media that works on surface filtration principle. As a result, the particles retention set in the case of surface filtration is much more restrictive than in the case of depth filtration.

Table 1. Comparative analysis of filter media

Depth filtration		Surface filtration	
advantages	disadvantages	advantages	disadvantages
low cost	hard to clean	possibility of cleaning and reuse	high cost
high efficiency	filter medium particles migration is possible	filter medium particles migration is excluded	low initial efficiency
high capacity to retain impurities	filtering performance is depending on the fluid viscosity	fatigue resistance, resistance to temperature or corrosive agents	limited capacity to retain impurities
	relatively large pressure drop	lower pressure drop	
	increased clogging	reduced clogging	

2. ANALYSIS OF WOVEN FILTER MEDIA FUNCTIONALITY

Woven filter media are products that are differentiated by structure and properties in strict accordance with the requirements and particularities of the process in which they operate. The filter medium structure is necessarily associated with the principle used to separate the mixture particles (surface filtration or depth filtration).

2.1. Features Woven Filter Media

The result of filter medium different properties combination sets up its quality and respectively its functionality. For an objective assessment of filter media quality (functionality) three groups of properties have been identified as follows:

- properties related to filter medium mounting system type. Those properties are important for the mechanical implementation of the filter respectively the filter medium set up on the support frame. Among the key properties of this group stated: stiffness, tensile strength, tear resistance, burst strength, abrasion resistance, vibration stability, elongation, the edges stability;

- properties related to the application type that are taking in consideration the compatibility between the filter medium and the processed medium. In this category falls the following properties: chemical stability, thermal stability, biological stability, dynamic stability, adsorption, absorption, operational safety and security, electrostatic characteristics, reuse capability, price;

- properties addressing specific filtrations process particularities underlining the filter medium capacity to comply with required demands. The most important properties of this group are: the smallest particle retained, retention efficiency, the structure of filter media, particle shape, filtering mechanisms used, flow resistance, porosity of filter media, permeability, tendency to clog, filter-cake discharge characteristics.

For each filter medium, depending on field of use and the requirements in service only some of these properties are necessary. As a result, the design of woven textiles intended to be used as filter media must be made in accordance with functionality criteria ensuring priority to the properties requested by the process utilized.

The relation structure – properties – use value is the design criterion for woven filters. Value engineering is a method of research and systemic design according to which the functions of the product studied (filter medium) must be designed and carried out with minimum expenditure in terms of highest quality, reliability and performance (Condurache et al., 2004). Value engineering instrumentation methodology implies the following stages:

- functional analysis: answers the questions what is and what the product does; the function list of the analyzed product is completed;
- classification of functions: answers the question how important the function is and how well meets the user requirements; function's relative importance, intrinsic and technical dimension terms are ascertained; functions classification for the analyzed product is finalized;

Table 2. The functions of woven filter media

Symbol	Function name	Technical dimension	Function type
F1	to separate the phases of a heterogeneous mixture	porosity permeability	primary, objective, necessary, general
F2	to ensure filtration fineness	shape, size and pore distribution	primary, objective, necessary, specific to filter principles
F3	to be dimensionally stable during operation	elastic and residual tensile strain	primary, objective, necessary, specific to filter principles
F4	to withstand the action of mechanical factors during operation	tensile strength burst resistance	primary, objective, necessary, specific to filtration process
F5	to withstand the erosive effects of the environment	chemical resistance	primary, objective, necessary, specific to filtered fluid
F6	to ensure filtration velocity	active filtration surface	primary, objective, necessary, specific to filter principle
F7	to withstand the erosive action of the filtered fluid	abrasion resistance	secondary, objective, necessary, specific to filtration process
F8	mechanical durability	fatigue resistance	secondary, objective, necessary, specific to filter type
F9	shouldn't clog	filter structure, pores shape	secondary, objective, necessary, specific to filter type
F10	easy to clean and rebuild	filter structure raw material	secondary, objective, necessary, specific to filter type
F11	easy to fabricate	filter structure, fabrication technology	secondary, objective, necessary, specific to filter type
F12	easy to install and replace	filter shape and dimensions	secondary, objective, necessary, specific to filter type

- product design or redesign based on required functions.
- Function is considered an essential attribute of the studied product expressed in terms of medium and user. In the same time, the function can be regarded as a characteristic of the product that determines a particular utility. The list of function classification is the starting point of value engineering studies.
- Product functions are determined by importance, the measurement method, its contribution to achieving use value and the degree of generality.
- Drawing up the list of filter medium functions is based on defining the filter medium and the conditions under which it works. In principle woven filters are intended to be used in filtration.

Based on such considerations the functions considered to be necessary for the filter medium, their technical elements of assessment and their typology classified by standard criteria are shown in Table 2 (I.Cioară&L.Cioară, 2009).

2.2. The Hierarchy Functions Woven Fabrics Filter

After developing the list of all functions their classification is done in order to establish the importance and weightiness of each function in rapport to all functions the product offers. The classification of the function is done using Value Engineering methods such as the Expertise Method or Imposed Decision Method. Regardless of the method used the classification is done considering all filter media functions or, selectively, group of functions, classified according to their typology.

Imposed Decision Method presents a high degree of objectivity (Condurache et al., 2004). To apply this method the following steps must be achieved: comparing the functions in pairs, calculating the importance coefficient for each function and classifying the functions by their importance coefficient value. By comparing the environmental functions as a filter to obtain decisions (0-1), (0.5-0.5) or (1-0). Scoring are considered: 0 considered less important function, 1 for the function considered more important, compared to 0.5 when the functions are valued as important. D total number of decisions resulting from the comparison of the n features of the filter is calculated with:

$$D = C_n^2 = \frac{n \cdot (n-1)}{2} \qquad (1)$$

I importance factor for each sample is calculated function the relationship:

$$I = \frac{N}{D} \qquad (2)$$

- where: N is the sum of points awarded;
- D - total number of decisions.

The filter media 12 functions obtained by weaving defined in Table 2 were divided into two groups: 6 primary and 6 secondary functions. Apply for group relationship of the main functions, which will be used to design, to establish the number of decisions as follows:

$$D = C_6^2 = \frac{6 \cdot (6-1)}{2} = 15 \qquad (3)$$

Table 3. Coefficient calculation Ranking

Function	Decisions															N	I
	1	2	3	4	5	6	7	8	9	10	11	12	13	14	15		
F1	0.5	0.5	1	1	0.5											3.5	0.233
F2	0.5					0.5	1	1	0.5							3.5	0.233
F3		0.5				0.5				0.5	0.5	0.5				2.5	0.166
F4			0				0			0.5			0.5	0		1	0.066
F5				0				0			0.5		0.5		0	1	0.066
F6					0.5				0.5			0.5		1	1	3.5	0.233

In Table 3 are comparative analysis, two by two principal functions. The last column of the table are shown the importance scores and values of each corresponding functions. Based on the values of the coefficients of importance to obtain the hierarchy of the main functions in the following sequence: F1, F2, F6, F3, F4, F5.

Priority functions set out in this way are taken into consideration when designing or redesigning filter media weaved in accordance with functionality criteria. In this respect the assessing criteria of woven filter media priority functions are summarized in Table 4.

The parameters specific to the woven filter with simple structure that will provide functional design criteria are: the relative porosity, pore shape and size, the pore distribution, the active filtering surface.

Table 4. Priority functions of the filter media and their assessment criteria

The name of the function	Specific technical dimension	Structural characteristics of fabric
to separate the phases of a heterogeneous mixture	pore size	fineness and density of yarns
to ensure filtration fineness	pore shape and distribution, filter medium fineness	fineness and density of yarns, weave
to ensure filtering velocity	adequate filtering active area	fineness and density of yarns
to be dimensionally stable during operation	structural and mechanical characteristics of yarn and fabric	the mechanical characteristics of the of yarns
to withstand the action of mechanical factors during operation	tensile strength burst resistance	the mechanical characteristics of the of yarns
to withstand the erosive effects of the environment	chemical resistance	the nature of raw material

The structural characteristics of woven fabric,which determine the parameters of filter fabric are: yarns count, thread density and weave.

3. STRUCTURAL AND FUNCTIONAL CHARACTERISTICS OF THE FABRIC FILTERS WITH SIMPLE STRUCTURE

Characterization and use of the fabric estimating filters with simple structure can be made by means of specific structural and functional characteristics (Behera, 2010; Cioară, 2002). Among these characteristics are mentioned: porosity, pore size and architecture, active filtering surface environment and filter fineness.

3.1. Porosity

Porosity, feature size filter material is the property of having pores in their structure (Medar&Ionescu, 1986; Cay et al., 2005). In connection with the porosity are two distinct notions:

- relative porosity, apparent or open, when taking into account only pores that communicate with each other;
- absolute porosity, effective or real, if we take into account all the pores, i.e. those who are isolated.
- Fluid flow through uniform or uneven spaces created by the filter medium, while maintaining the quality of filtration, filtration efficiency and smoothness and filtering capacity are issues directly related to the porosity of filter media.
- Fluid movement across the filter medium is described by the filtration rate, defined as the maximum volume of fluid passing per unit time through unit area of filter. Porosity refers to the filter media pore volume per unit volume and is typically seen in relative units. Generally, the textile filtering media are inhomogeneous because the filter permeability changes during the exploitation. The medium in homogeneity can be bigger or smaller, depending on the structure of woven filter.

3.2. Pore Dimensions and Architecture

An important feature of each filter surfaces is the existence of pores which penetrate the entire thickness of the filter and retain solid particles larger than the pores in the cross section of their most narrow, but allow passage of fluid that carried them. Small pore is a void within a solid body. After dimensions are distinguished (Medar&Ionescu, 1986) : fine pores with a diameter greater than 20 μm (invisible to the naked eye) and coarse pore diameter greater than 20 μm (visible to the naked eye). The way of communication with the outside pores can be:

- open, when communication with the outside;
- closed, when no communication with the outside.
- Dimensional uniformity and stability of pores of a filter medium directly influences the process of filtering performance (Gabrijelcic et al., 2009). Pore size and shape of woven textile filter media are dependent on the basic structural parameters of fabric: the fineness of yarns, thread density and the weave.
- Pore's characteristics which is assessed functional performance of a fabric filter are: side pore, pore area, architecture and distribution of pores in the fabric plane.
- Side and pore area are geometric features of woven fabric due to its basic structural parameters.
- In terms of basic structural parameters, the woven fabrics with simple structure can be balanced or unbalanced in fineness and density yarns, and the resulting pores have square or rectangular form.

In Figure 2 are defined pore geometry of a woven fabric with simple structure. In the balanced woven fabric (Figure 2.a), the warp and weft, have the same fineness and density, the same diameter d and the same density P. Therefore to obtain a filter fabric with square pores. In the unbalanced woven fabric (Figure 2.b), the warp and weft have different count, expressed by d_u, d_b, and different density threads expressed by Pu, Pb. As a consequence to obtain a fabric filter with rectangular pores.

3.2.1. Pore Side

Pore side is the distance between two consecutive threads of the fabric measured in the projection on the horizontal fabric's plan.

For structures balanced (Figure 2.a), pore side it could be calculated with:

$$l = \frac{10}{P} - d \ (\text{mm}) \tag{4}$$

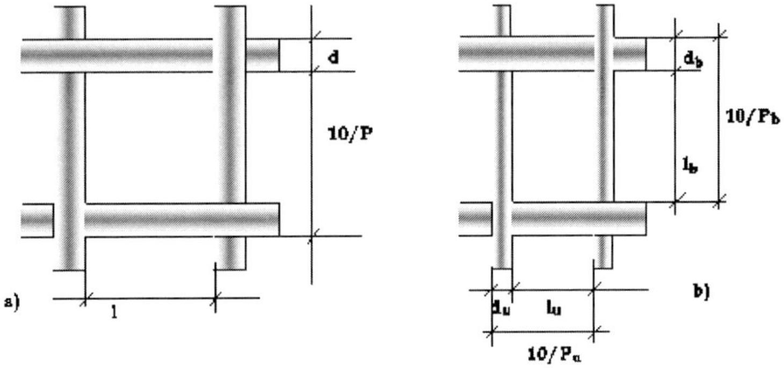

Figure 2. Pore geometry

For structures unbalanced (Figure 2.b), pore sides, l_u and l_b, is defined by relations:

$$l_u = \frac{10}{P_u} - d_u \ (\text{mm}); \ l_b = \frac{10}{P_b} - d_b \ (\text{mm}) \tag{5}$$

3.2.2. Pore Area

Pore area is defined as the projection on the horizontal plan of the fabric's pore.

For balanced structure (Figure 2.a) the pore area A_p is calculated with:

$$A_p = l^2 \ (\text{mm}^2) \tag{6}$$

For unbalanced structure (Figure 2.b) pore area is calculated by the relationship:

$$A_p = l_u \cdot l_b \ (\mathrm{mm}^2) \tag{7}$$

3.2.3. Pores Architecture

Pores architecture is a characteristic determined by the fineness of yarns, threads density, mechanical and rheological characteristics of yarns and weave used. The segments of the yarn which constitute the pore sides can be considered, like a beam in one of two situations:

- passing from one side to another fabric,
- above or under the opposite yarn system.
- Under these circumstance, the pores` shape and size depend not only on yarn count and density, but also on the positions of the yarns in the weave.

For this purpose an analysis is presented which highlights the fact that the weave determines the distribution of requests in the yarns and the default form of pores. The analysis is done on three woven fabrics filter, whose characteristics are presented in Table 5.

The weave used in implementing the three fabrics are shown in Figure 3. To the right of each weave are represented by pores with a distinct architectural (Cioara et al., 2003).

Table 5. Variants of filter fabrics

Woven filter	Weave	Raw material	Yarn diameter (mm)	Thread density (yarns/cm)
Filter 50 mesh	Plain	Polyamide	0.14	20
Filter 22 mesh	Twill D2/2	Polyamide	0.45	9
Filter 24 mesh	Twill D3/1	Polyamide	0.45	9.5

After examining the shape of pores in the three weave can be made the following observations and interpretations:

- at the fabric filter with plain weave (Figure 3.a) all the pores have the same architecture; the threads have a similar position in the pore sides (all threads are crossing from one side to another of the fabric). Under these conditions the fabric structure creates the potential formation of uniform pores in the shape of their; the pores I, is identical in structure with pores II;

- at the fabric filter with twill weave D 2/2 (Figure 3.b) pores of the report have the same structure. Shapes and sizes of the four types of pores are identical. It creates the conditions to achieve a uniform structure with a high degree of homogeneity to ensure a quality filter; pores numbered I, II III IV is identical in structure;

- at the fabric filter with twill weave D 1/3 (Figure 3.c) is classified in four types of pores. The four distinct architecture creates pores with different shapes and volumes that the conditions for the flow through it are differentiated; pores numbered II, is identical in structure with pores IV, pores I and III is different.

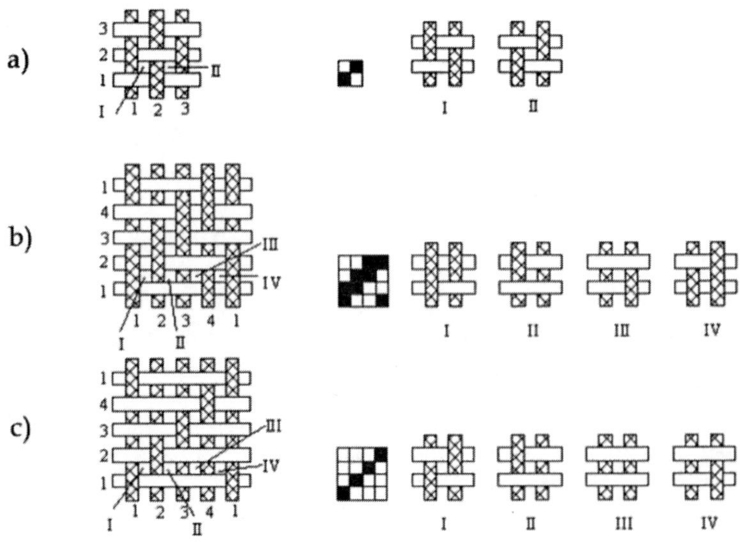

Figure 3. Type of pores

To support these interpretations, on the three types of fabric filter were performed measurements of area pores, using a specialized program. Measurements were made on the fabric images captured with a stereo microscope with digital camera.

These images are presented in Figure 4, 5 and 6. In each figure is shown a fragment of the microscopic image of the fabric filter with the contours measured pore, the statistics of individual value of area and pore distribution curve.

To facilitate analysis of information provided by research, in Table 6 were centralized statistical processing on the string values of individual values of the areas measured.

Table 6. Statistical evidence

Weave	Mean area (mm^2)	Min. value (mm^2)	Max. value (mm^2)	Range (mm)	Std.Dev. (mm)	CV (%)
Plain	0.1021164	0.0841335	0.1120637	0.0279302	0.0064476	6.31
Twill D2/2	0.2104919	0.1555840	0.2731934	0.1176094	0.0310180	14.73
Twill D3/1	0.3233822	0.1810607	0.4755648	0.2945041	0.0722753	22,34

The analysis of microscopic images shows that the filter fabric with plain weave (Figure 4) pores are relatively uniform shapes and sizes. This is supported on the one hand, the low dispersion of individual values (s=0.00644) and, on the other hand, the restricted distribution of individual values around the average. Extreme values, minimum and maximum, with reduced weight, have a deviation of up to 10% of the average pore area.

At the fabric filter with twill weave D 2/2 (Figure 5) is observed as architecture, two types of pores with greater irregularity than plain weave. Pore area shows a greater variation, which is confirmed by the dispersion value (s=0.03101) and the pore distribution curve shape. Even if the pore area varies widely, up to 25% from the mean, the woven filter is estimated that the structure is uniform. Extreme values are numerous and, consequently, the distribution curve is wider.

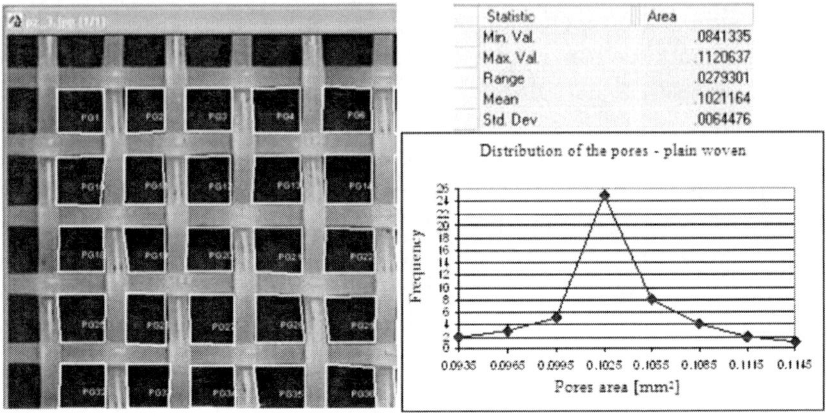

Figure 4. Pore architecture of the woven filter - plain weave

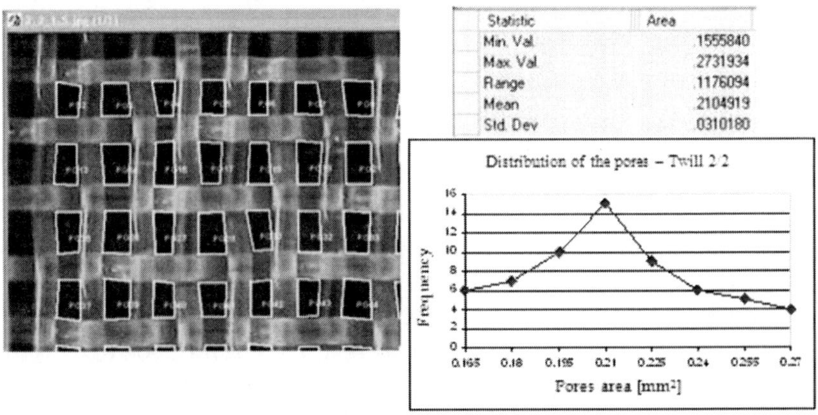

Figure 5. Pore architecture of the woven filter - twill D 2/2

At the fabric filter with twill weave D 3/1 (Figure 6) is observed as architecture, the four specific types of pores. The dispersion of measured values (s=0.07227) indicates large variation in pore area, which is emphasized by the distribution curve. Pore area varies widely, with over 50% of the mean value and the structure of a fabric filter has emphasized the uneven character.

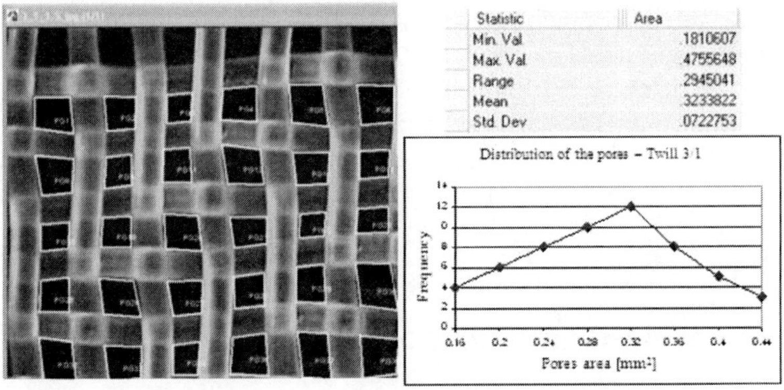

Figure 6. Pore architecture of the woven filter - twill D 3/1

3.3. The Active Filtering Surface

This parameter provides information about the porosity of the woven filter and resistance to fluid flow through it. Active filtering area was defined as the ratio of pore area, A_p, and area of fabric element, A_{et}, is calculated from the relationship:

$$F_d = P_u P_b \left(\text{pores} / \text{cm}^2\right) \tag{8}$$

3.4. Filter Fineness

The woven filter fineness is a nominal identification, which is expressed by the number of pores per unit length or number of pores per unit area. The filter fineness is calculated by the following relations:

4. ALGORITHMS FOR FUNCTIONAL DESIGN OF FILTER. EXAMPLES OF APPLICATION

The last stage of the value engineering technique is to design or redesign based priority functions. Based on the conclusions presented in the above analysis were developed five functional design algorithms for simple structure filter fabrics balanced and unbalanced in fineness and/or density (I.Cioara & L.Cioara, 2010). Always choose the algorithm is done according to known elements (input data) and the requirements process of filtering (output data) to design, redesign or verification. Following the sequence of calculation steps is different from one algorithm to another.

4.1. Algorithms For Functional Design Of Filter

In Tables 7 and 8 shows the deployment of the calculations according to data input and output, these calculations are completed to the mass of fabric filter.

In the relationships of calculating the parameters of qualitatively appraising the simple structure woven fabrics used as filtering media the significance of the employed symbols is the following: d - yarn diameter in the balanced structures (mm); d_u, d_b - warp and weft diameter (mm); A - tabled constant for the diameter calculating; A_p - pore area, (mm^2); A_{et} - woven fabric element area, (mm^2); T_{tex} - yarn count in the balanced structures, (g/km); T_{texu}, T_{texb} - warp and weft count, (g/km) ; P - threads density in the balanced structures, (yarns/cm); P_u, P_b - warp density and weft density, (yarns/cm); l - square pore side, (mm); l_u, l_b - the pore side in the warp and weft direction, (mm); F - balanced structure filter fineness, (pores/cm); F_m - balanced structure filter fineness, (pores/inch); F_d - filter fineness, (pores/cm^2); M - woven fabric mass, (g/m^2); a - crimp yarn in the woven fabric, (%); m – factor of unbalanced for threads density.

4.1.1. Algorithm I. Design Of Simple Filters Based On The Density And Fineness Of Yarn (Table 7)

In this case we consider as known the basic structural characteristics of fabric: warp and weft yarn diameters (d_u, d_b) and their densities (P_u, P_b). The algorithm is used to identify specific characteristics of woven fabrics filters: the filter fineness, the pore side and area, the active filtering surface.

Based on of factors identified can be appreciated if appropriate filter fabric filtration process characteristics. Depending on the conclusions formulated after verification can be redesign the fabric to match the outlays.

4.1.2. Algorithm Ii. Design Of Simple Filters Based On The Yarn Diameter And Filter Fineness (Table 7)

In this case it is considered known diameters of the warp and weft of yarns (d_u, d_b) and fineness filter fabrics. The algorithm is used to design a filter of woven fabrics density unbalance. The yarns density, the pores sides and surface, the active filter area and the fabric weight are determined by calculations.

Since by changing the density unbalance factor the pores shape and dimensions are also modified ensuring an appropriate filtering fineness. The algorithm can be additionally used to re-design according to given requirements.

4.1.3. Algorithm Iii. Design Of Simple Filters Based On The Yarn Diameter And Side Of Pore (Table 7)

The algorithm is used to design the fabric characteristics in compliance with the parameters of the filtering process. The input data are warp and weft yarn diameters (d_u, d_b) and pore sizes (l_u, l_b). Imposing pore side means the particle to be retained is known, the filter being a calibrated restriction. The required yarn density and all the specific characteristics of the filter are obtained by calculations.

4.1.4. Algorithm Iv. Design Of Simple Filters Based On The Yarn Diameter And Active Filter Area (Table 8)

Active filter area is a feature that determines the permeability of the fabric and filter potential to withstand the fluid flow. This algorithm is considered known following elements: the warp and weft yarn diameters (d_u, d_b) and active filtering surface (S_a).

Table 7. Specific elements for algorithms design I, II, III

Algorithm I	Algorithm II	Algorithm III
1. Input data: d_u, d_b, P_u, P_b	1. Input data: d_u, d_b, F_d	1. Input data: d_u, d_b, l_u, l_b
2. Yarn count $T_{tex} = \frac{d^2}{A^2}$	2. Yarn count $T_{tex} = \frac{d^2}{A^2}$	2. Yarn count $T_{tex} = \frac{d^2}{A^2}$
3. Filter fineness balanced structures $F = P_u = P_b$ *pores / cm* $F_m = F \cdot 2.54$ pores/inch unbalanced structures $F_d = P_u \cdot P_b$ pores/cm²	3. Thread density $P = f(F_d)$ $P_u \cdot P_b = F_d$ $\frac{P_u}{P_b} = m$ $P_u = \sqrt{F_d \cdot m}$ $P_b = \frac{P_u}{m}$	3. Thread density $P = f(l_p, d)$ $P_u = \frac{10}{l_u + d_u}$ $P_b = \frac{10}{l_b + d_b}$
4. Pore side $l_u = \frac{10}{P_u} - d_u$ $l_b = \frac{10}{P_b} - d_b$	4. Pore side $l = f(P, d)$ $l_u = \frac{10}{P_u} - d_u$ $l_b = \frac{10}{P_b} - d_b$	4. Filter fineness balanced structures $F = P_u = P_b$ *pores / cm* $F_m = F \cdot 2.54$ pores/inch unbalanced structures $F_d = P_u \cdot P_b$ pores/cm²
5. Pore area $A_p = f(P, d)$ $A_p = \left(\frac{10}{P_u} - d_u\right) \cdot \left(\frac{10}{P_b} - d_b\right)$		
6. Filtering active surface $S_a = f(P, d)$ $S_a = (10 - P_u \cdot d_u) \cdot (10 - P_b \cdot d_b)$		
7. Woven fabric mass $M = \frac{P_u \cdot T_{texu}}{10} \cdot \frac{100}{100 - a} + \frac{P_b \cdot T_{texb}}{10} \cdot \frac{100}{100 - a}$		

The algorithm allows the calculation of specific characteristics of the filter: pore area and pore side, which determines the filtration fineness and efficiency.

4.1.5. Algorithm V. Design Of Simple Filters Based On The Pore Side And Active Filter Area (Table 8)

In this case filter fabrics design is based on filtering process requirements. Input data are: pore sizes (l_u, l_b) and active filtering surface (S_a). Active filter area determines the filter permeability and influences the flow velocity. The pore side determines the size of the retained particle and influences the filtration fineness and efficiency. The structural characteristics of the fabric are obtained by calculations.

4.2. Examples Of Application

Further gives examples of the implementation of the five algorithms to design filter woven fabrics with simple structure made from monofilament yarns.

To facilitate the calculation has been made a computer assisted design program (Cioara et al., 2008).

Sequence with the principal program menu is shown in Figure 7. At the top of the screen, interactive buttons, the corresponding input data are highlighted five algorithms. The dates presented in Figure 7are typical application example of the algorithm I (Table 9).

The program has the possibility of saving data Input and account for the different variants. These data are provided to the operator in the form of tables. In this way the designer can to compare specific elements of the different variants and select the variant that best corresponds to requirements process of filtering.

The applications are presented in Tables 9, 10, 11, 12, 13. These applications are made for woven fabrics filters with finesse and densities balanced structures, made with weave plain, weave – hopsack 2/2, weave-irregular sateen and weave – twill 2/2.

Each table is presented as a fabric filter microscopic image and the list of parameters supplied by the computer program. Same time, the tables are highlighted input data specific to each algorithm separately.

Table 8. Specific elements for algorithms design IV, V

Algorithm IV	Algorithm V
1. Input data: d_u, d_b, S_a	1. Input data: l_u, l_b, S_a
2. Yarns count $T_{tex} = \frac{d^2}{A^2}$	2. Pore area $A_p = l_u \cdot l_b$
3. Thread density for balanced structures of cover factory: $P_u \cdot d_u = P_b \cdot d_b$ $P = f(S_a, d)$ $P_u = \frac{10 - \sqrt{S_a}}{d_u}$ $P_b = \frac{10 - \sqrt{S_a}}{d_b}$	3. Thread density $P = f(S_a, l_p)$ $S_a = \frac{A_p}{A_w} \cdot 100 \Rightarrow A_w = \frac{A_p \cdot 100}{S_a} = \frac{10}{P_u} \cdot \frac{10}{P_b}$ $P_u \neq P_b$; $\frac{P_u}{P_b} = m$ $P_b = \sqrt{\frac{S_a}{m \cdot A_p}}$ $P_u = P_b \cdot m$
4. Filter fineness balanced structures $F = P_u = P_b$ *pores / cm* $F_m = F \cdot 2.54$ pores/inch unbalanced structures $F_d = P_u \cdot P_b$ pores/cm^2	4. Filter fineness balanced structures $F = P_u = P_b$ *pores / cm* $F_m = F \cdot 2.54$ pores /inch unbalanced structures $F_d = P_u \cdot P_b$ pores/cm^2
5. Pore side $l = f(P, d)$ $l_u = \frac{10}{P_u} - d_u$ $l_b = \frac{10}{P_b} - d_b$	5. Yarns diameter $d_u = \frac{10}{P_u} - l_u$ $d_b = \frac{10}{P_b} - l_b$
6. Pore area $A_p = f(P, d)$ $A_p = \left(\frac{10}{P_u} - d_u\right) \cdot \left(\frac{10}{P_b} - d_b\right)$	6. Yarns count $T_{tex} = \frac{d^2}{A^2}$
7. Woven fabric mass $M = \frac{P_u \cdot T_{texu}}{10} \cdot \frac{100}{100 - a} + \frac{P_b \cdot T_{texb}}{10} \cdot \frac{100}{100 - a}$	

In each example there is a good correlation between the elements provided by the computer program and structural characteristics shown on the microscopic image of the fabric made. This aspect allows us to say that the proposed algorithms achieve a better modeling of the structural parameters and specific functional of woven fabrics filters with simple structure.

Table 9. Balanced structure - designed with algorithm I (weave plain)

Sample filter	Parameter	Symbol	Value	UM
	Yarns diameter	d	**0.31**	mm
	Thread density	P	**11.0**	threads/cm
	Filter finess	F	11.0	pores/cm
	Filter finess mesh	F_m	27.94	pores/inch
	Pore side	l	0.599	mm
	Pore area	A_p	0.3589	mm^2
	Filtering active surface	S_a	43.43	%
	Dimensional factor	A	**0.0392**	-
	Yarns count	T	62.54	tex
	Crimp yarn	a	4	%
	Woven fabric mass	M	143.32	g/m^2

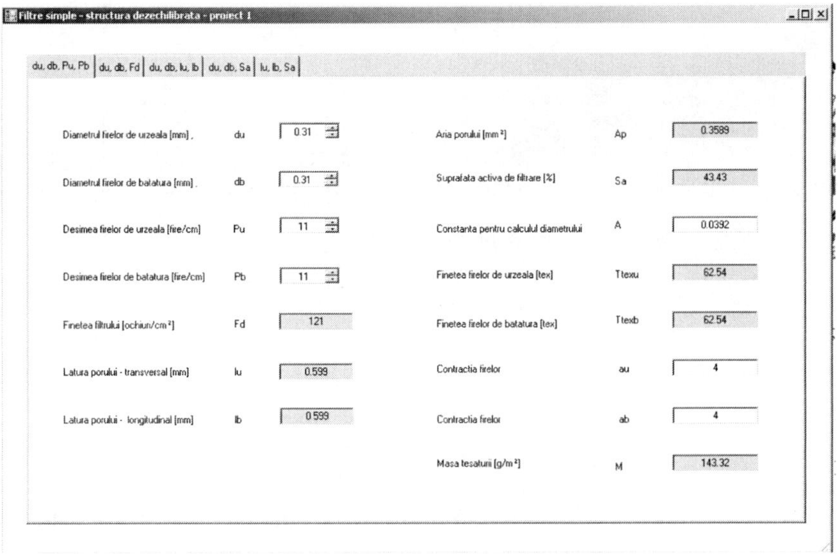

Figure 7. The main menu of the program

Table 10. Balanced structure - designed with algorithm II (weave – hopsack 2/2)

Sample filter	Parameter	Symbol	Value	UM
	Yarns diameter	d	**0.29**	mm
	Filter finess mesh	Fm	36	pores/inch
	Filter finess	F	14.2	pores/cm
	Thread density	P	14.2	threads/cm
	Pore side	l	0.414	mm
	Pore area	A_p	0.1716	mm^2
	Filtering active surface	S_a	34.6	%
	Dimensional factor	A	**0.0392**	-
	Yarns count	T	54.73	tex
	Crimp yarn	a	2	%
	Woven fabric mass	M	158.61	g/m^2

Table 11. Balanced structure - designed with algorithm III (weave- irregular sateen)

Sample filter	Parameter	Symbol	Value	UM
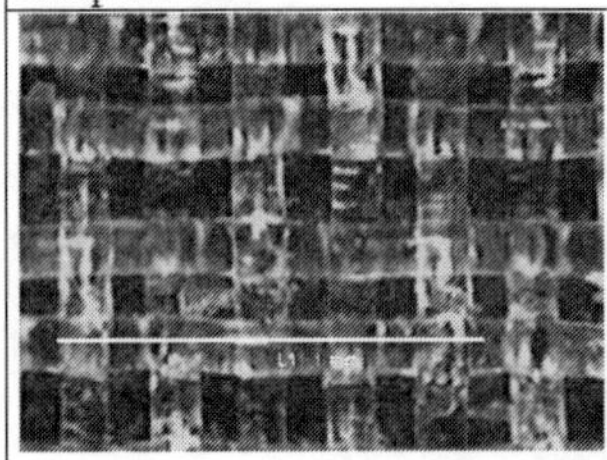	Yarns diameter	d	**0.11**	mm
	Pore side	l	**0.11**	mm
	Thread density	P	45.5	threads/cm
	Filter finess	F	45.5	pores/cm
	Filter finess mesh	Fm	115.57	pores/inch
	Pore area	Ap	0.0121	mm^2
	Filtering active surface	Sa	24.95	%
	Dimensional factor	A	**0.0392**	-
	Yarns count	T	7.87	tex
	Crimp yarn	a	4	%
	Woven fabric mass	M	74.6	g/m^2

5. CONCLUSION

The woven fabrics which are used as filter fabrics, have the functionality imposed for the filtration process. Structure and properties of the woven filter fabrics are adequately differentiated, for the principles of the filtering process. The paper defines the structural and functional elements that are specific to the filtering woven fabrics, which have a structure that is simple, balanced and unbalanced in yarn count and thread density.

Table 12. Balanced structure - designed with algorithm IV (weave – plain)

Sample filter	Parameter	Symbol	Value	UM
	Yarns diameter	d	**0.14**	mm
	Filtering active surface	S_a	**49**	%
	Thread density	P	21.4	threads/cm
	Filter finess	F	21.4	pores/cm
	Filter finess mesh	F_m	54	pores/inch
	Pore side	l	0.327	mm
	Pore area	A_p	0.107	mm^2
	Dimensional factor	A	**0.0392**	-
	Yarns count	T	12.76	tex
	Crimp yarn	a	5	%
	Woven fabric mass	M	57.49	g/m^2

Table 13. Balanced structure - designed with algorithm V (weave – twill 2/2)

Sample filter	Parameter	Symbol	Value	UM
	Pore side	l	0.1	mm
	Filtering active surface	S_a	**31**	%
	Pore area	A_p	0.01	mm^2
	Thread density	P	55.7	threads/cm
	Filter finess	F	55.7	pores/cm
	Filter finess mesh	F_m	141.4	pores/inch
	Yarns diameter	d	0.08	mm
	Dimensional factor	A	**0.0392**	-
	Yarns count	T	4.16	tex
	Crimp yarn	a	6	%
	Woven fabric mass	M	49.3	g/m^2

The methods of filter design for the fabrics with simple structure is based on the specific geometry of the structure elements fabrics. To achieve filter fabric with uniform pore (size and shape), is recommended of use, weaves balanced with equal segments and uniform distribution (plain, twill D 2/2, hopsack 2/2). Five algorithms of designing the filtering woven fabrics having a simple structure were elaborated, which can be differentially applied depending on the initially introduced elements.

By using computer aided design program to obtain a set of variations of the designer select the optimal variant, which corresponds better to the needs of field exploitation.

ACKNOWLEDGEMENTS

The paper was developed in the research program PNCDI II no.31-088/2007 contract, financed by the Romanian government, which the Technical University "Gheorghe Asachi" of Iassy was a partner. Research presented in this work were performed at the Faculty of Textiles and Leather Engineering of Iassy, Department Technology and Design of Textile Products, laboratory TEXTILEXPERT.

REFERENCES

1. S. Adanur, (1995, (1995).Handbook of Industrial Textiles, Technomic Publishing Company, 1-56676-340-1 U.S.A.

2. Behera, B.K.&Hari,P.K. (2010).Woven textile structure, Woodhead Publishing Limited, ISBN 978-1-84569-514-9, Cambridge, UK

3. A. Cay, S. Vassiliadis, M. Rangoussi, I. Tarakcioglu, (2005, (2005).On the use of image processing techniques for the estimation of the porosity of textile fabrics, World Academy of Science, Engineering and Technology, 2 2 76 79 , 2010-3778

4. I. Cioară, L. Cioară, (2010, (2010).Algorithms design of the filtering woven fabrics with simple structure, Annals of the University of Oradea- fascicle of Textiles Leatherwork, volume XI, 41 46 , 1582-5590 Oradea, Romania

5. I. Cioară, L. Cioară, (2009, (2009).Criteria for estimating the functionality of the filtering mediaobtained by weaving, Industria Textilă, 1 21 25 , 1222-5347 Bucharest

6. I. Cioară, L. Cioară, D. Cascaval, (2008, (2008).Method and software design woven filter media, The VI National Conference of Virtual Learning (CNIV), University of Bucharest

7. pp.60 6. Publishing, I. S. S. N. 1842-470, Constanţa, România

8. I. Cioară, L. Cioară, E. Onofrei, 2003 Aspects regarding the architecture of the filtering woven fabrics, UNITEX, 1 28 30 , Belgium

9. L. Cioară, (2002, fabric. Woven, Performantica. structure, House. I. S. B. N. Publishing, 973-8075-1, 1. , Iassy, România

10. L. Cioară, I. Cioară, 2001 Optimization of the technical parameters in processes of filter weaving, Proceedings vol.I, 1st Autex Conference Technitex-Technical Textiles: Designing Textiles for Technical Applications, 9-72984-683-9 467 470 , Minho

11. L. Cioară, I. Cioară, O. Marchiş, 1991 The influence of certain technical parameters onthe capacity of filtering the woven textile planes, Industria Uşoară, 4 239 241 , 1222-5347 Bucharest, România

12. G. Condurache, R. Ciobanu, M. Niculae, (2004, (2004).Analysis and value engineering, Performantica Publishing House, 973-730-022-x Iassy, România

13. H. Gabrijelcic, R. Urbas, F. Sluga, K. Dimitrovski, (2009, (2009).Influence of Fabric Constructional Parameters and Thread Colour on UV Radiation Pretection, FIBRES&TEXTILES in Eastern Europe, 17 17 1 72), 46 54 , 1230-3666 Lodz, Poland

14. A. R. Harracks, S. C. Anand, 2000 Handbook of technical textiles, Woodhead Publishing Limited, 1-85573-385-4

15. O. Marchiş, L. Cioară, I. Cioară, (1991, (1991).Considerations on certain factors that influence the porosity of filtering planes, Industria Uşoară, 4 4 237 238 , 1222-5347 Bucharest, România

16. S. Medar, F. Ionescu, 1986 Filters for hydraulic and pneumatic drives, Technical Publishing House, Bucharest, România

17. H. K. Rouette, 2001 Encyclopedia of Textile Finishing, Woodhead Publishing Limited, 1-84569-065-6

Chapter 8

Studies on the mechanical properties of woven jute fabric reinforced poly (L-lactic acid) composites

G.M. Arifuzzaman Khan[a], M. Terano[c], M.A. Gafur[b], M. Shamsul Alam[a]

[a] Department of Applied Chemistry and Chemical Technology, Islamic University, Kushtia 7003, Bangladesh

[b] PP and PDC Division, BCSIR, Dhaka 1205, Bangladesh

[c] School of Materials Science, Japan Advanced Institute of Science and Technology, Japan

ABSTRACT

Development of ecofriendly biocomposites to replace non-biodegradable synthetic fiber composites is the main objective of this study. To highlight the biocomposites as a perfect replacement, the plain woven jute fabric (WJF) reinforced poly(L-lactic acid) (PLLA) composites were prepared by the hot press molding method. The influence of woven structure and direction on the mechanical properties i.e. tensile, flexural and impact properties was investigated. The average tensile strength (TS), tensile modulus (TM), flexural strength (FS), flexural modulus (FM), and impact strength (IS) of untreated woven jute composite (in warp direction) were improved about 103%, 211%, 95.2%, 42.4% and 85.9%, respectively and strain at maximum tensile stress for composite samples was enhanced by 11.7%. It was also found that the strengths and modulus of composites in warp direction are higher than those in weft direction. WJF composites in warp and weft directions presented superior mechanical properties than non-woven jute fabric (NWJF) composites. Chemical treatment of jute fabric through benzoylation showed a positive effect on the properties of composites. Morphological studies by SEM demonstrated that better adhesion between the treated fabric and PLLA was achieved.

KEYWORDS

Woven jute fabric/PLLA biocomposite; Warp and weft directional jute fabric; Jute modification; Mechanical properties

1. INTRODUCTION

Solid waste disposal has become a burning problem nowadays. Garbage wastes such as plastic grocery bags, food packaging materials, bottles, containers etc. are mainly responsible for causing environment pollution in urban areas due to non-biodegradable nature. At the same time, the emission of green house gases during combustion is a serious global-scale problem. To keep our environment safe and green, it is compulsory to reduce the use of such pollution causing garbage. That is why, many countries banned plastic grocery bags responsible for so called "White Pollution". Alternative usage of biodegradable plastics and their natural fiber biocomposites gained more popularity because they are degraded easily after landfill. Therefore, it may lead to one of the solutions for the issue of disposal ground depletion.

Keeping this target, recent research efforts are being harnessed in developing a new class of fully biodegradable composites (biocomposites) by combining natural fiber with biodegradable polymer (Baghaei et al., 2013). A number of commercially available biodegradable polymers such as polyhydroxyalkanoate (PHA), poly(L-lactic acid) (PLLA), polycaprolactone (PCL), polyesteramide (PEA) etc. are used frequently as matrices of natural fiber composites. In the class of biodegradable polymers, PLLA has the greatest potential in the composite industries because of its facile availability, good biodegradability and high mechanical properties (Shukor et al., 2014). Though PLLA has superior properties to polyethyleneterephthalate (PET) in various applications, some drawbacks such as brittleness, inferior resistant on thermal deformation and lower water vapor barrier properties have limited its extensive use (Petersson et al., 2007). However, by incorporating natural fiber into the PLLA matrix can successfully improve the overall properties of composites.

Short natural fiber/PLA composites have been investigated by several studies (Oksman et al., 2003, Bax and Mussing, 2008, Sawpan et al., 2007 and Ganster et al., 2006). The average tensile strength found was about 50 MPa using compression molding and the injection molding technique. Modulus of short fiber reinforced composites was also found to be lower. The long hemp fiber/PLA composites were studied by Graupner et al. (2009). Mechanical strengths of the composite were also not so improved. On the other hand, woven fabrics have received extensive study as reinforcements since they provide acceptable mechanical properties, simple processing and conformability for advanced structural applications (Jekabsons and Bystrom, 2002). Woven fabric composites recommend more balanced properties than unidirectional laminas. The weaving of the fiber gives an interlocking that increases strength higher than fiber–matrix binding strength. Since fibers are attached together tightly in woven fabrics, pull-out is quite impossible, consequently need more strength to breakdown the fabric reinforced composites (Lomov et al., 2006). Sapuan et al. (2006) investigated the mechanical properties of woven banana fiber reinforced with epoxy composites. The tensile stress was found greater in X-direction (warp) than Y-direction (weft). Weaving patterns such as plain, twill, satin, basket, leno etc. are crucial factor in determining the response of the composites (Pothen et al., 2006).

A number of reports on WJF composites have been found so far but none deals with the use of WJF as reinforcement in poly(L-lactic acid). In most cases jute fiber was used either in warp or weft direction for making hybrid woven fabric composites with synthetic polymer matrix (Khalil et al., 2011). Reinforcing 2D and 3D woven structures for natural fabric reinforced polymer composites were developed by Lomov et al. (2009). However, the properties of plain WJF reinforced polyester composites were found to have tremendous potential for improving the performance of composite structures due to longer stability, higher tensile, flexural, in-plane shear, interlaminar shear, impact properties and low manufacturing cost of plain woven fabric (Ahmed and Vijayarangan, 2007). Therefore, plain woven jute fabrics have been introduced in this study to fabricate composite by the compression molding technique. In order to get good adhesion between fiber and matrix (which give better mechanical strength in natural fiber/fabric reinforced composites) fibers have been modified chemically as like Lai and Mariatti, 2008. The effect of fiber orientation (warp and weft) on mechanical properties of woven fabric composites has also been reported.

2. EXPERIMENTAL

2.1. Materials

The PLLA polymer having an average molecular weight of 110,000 g mol^{-1} was synthesized in melt polycondensation methods (Khan et al., 2013). Unbalanced plain weave jute mat 52 × 44 (52 yarns in warp direction and 44 yarns in weft direction per 100 mm), having an average weight 200 g m^{-2} and average thickness of 6 mm was collected from Fatima-Alyaf Tala-e Jute Industries Ltd. Gazipur, Bangladesh (Fig. 1). Raw jute (unidirectional mat) was collected from local market of Jhenidha, Bangladesh.

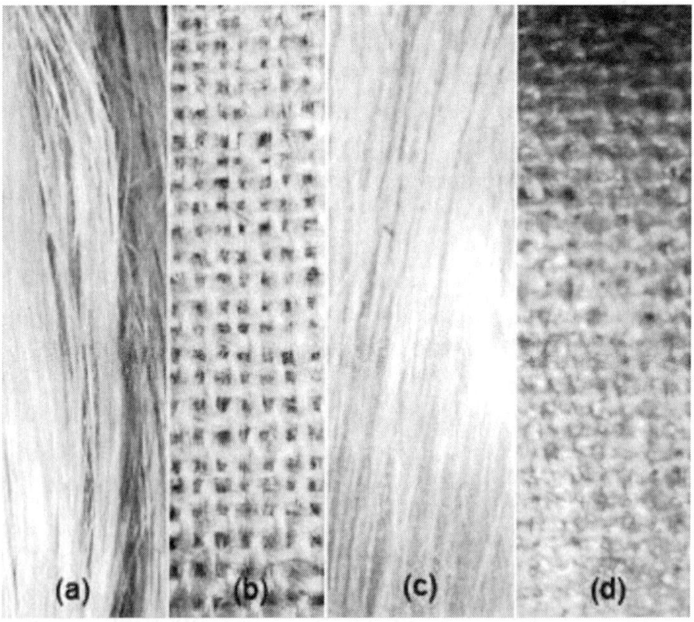

Figure 1. Image of (a) unidirectional jute, (b) plain woven jute fabric, (c) nonwoven jute fabric composite and (d) plain woven jute fabric composite.

2.2. Molding of woven and nonwoven fabric composites

Composite fabrication process was completed in two steps. In the first step PLLA films approximately 0.2–0.3 mm in thickness were prepared by 2 polished stainless steel (SS) sheets in a hot press molding machine. Molding machine is accomplished by hydrolic press and hot plate. Molding temperature and pressure can be as high as 250 °C and 250KN, respectively. SS sheets were kept for 10 min at 180 °C and 50KN pressure thereafter it was cooled by tap water. For the second step, square WJF of (15 × 15) cm^2 in dimensions were cut and dried in an oven at 60 °C for 24 h to remove moisture. The prepared PLLA films were also cut into the same dimension like fabric. A rectangular stainless steel (SS) mold with 15 × 15-cm square cavity and 10 mm depth was used for composite fabrication. The mold was packed by WJF in between the PLLA films maintaining fixed weight fraction of woven fabric in composite. A mold releasing spray (BONEY-Mould release silicon spray, manufacturer-London chemicals Ind. England) was used for easy opening the mold. The mold was then placed on a hot press molding machine and composite was fabricated as like PLLA film making. Same procedure was followed for NWJF/PLLA composites, where fibers were placed unidirectionally. Weight fractions of fabrics in composites were calculated by the following formula:

$$W_f = \frac{W_j}{W_m + W_j}$$

where W_f, W_j, and W_m are the weight fractions of fabric in composite, weight of jute fabric and weight of polymer matrix respectively.

2.3. Mechanical measurements

Tensile test was conducted in both warp and weft directions according to ASTM D882 (E) using a Universal Testing Machine (Hounsfield UTM 10KN). The specimens for tensile were cut in 30 mm wide dumbbell shape with an overall length of 110 mm and a 50 mm length of the narrow, 10 mm wide parallel-sided section. The dimension of clamped area between two clips was 40 mm × 15 mm × 0.5–1.0 mm. Test speed was fixed 2 mm/min. The yield stress, stain and modulus were measured by QMAT Pro for Textile Software Package.

Three-point flexural tests of composites were carried out using same Hounsfield UTM 10KN attached with computer program (QMAT Pro for Textile Software Package) according to the standard method used for flexural properties (ASTM D790–98). The specimens for tensile were cut into the dimensions: 79 mm × 10 mm × 0.5–1.0 mm. Test speed was fixed at 2 mm/min.

Notched Charpy impact tests (according to ASTM D6110–97) were carried out using a universal impact tester. A rectangular piece of sample was prepared according to Fig. 2. The specimen is having a V-notch at a point equidistant from the ends of the long side. All the results were taken as the average value of 10 samples. Unreinforced PLLA sample was also tested for comparison purpose.

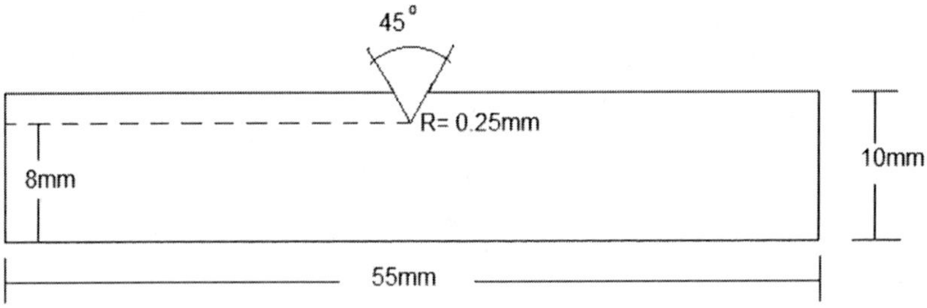

Figure 2. Simple beam notch Charpy impact test specimen.

2.4. Scanning electron microscopy

The tensile fractured surface morphology of composites was examined by scanning electron microscope (Philips XL-30). The instrument was operated with an accelerating voltage 30 kv. The samples were coated with 3 nm gold layer using a vacuum sputter coater.

3. RESULTS AND DISCUSSION

Figure 3 and Figure 4 represent the tensile stress–strain diagrams of unreinforced PLLA, untreated and treated WJF/PLLA composites in warp and weft direction, respectively. The initial portion of the curve of untreated WJF/PLLA composites is linear at low strain rates followed by a change in the slope of the curve indicating nonlinear behavior of the material. The start of nonlinearity in the curve is an indication of the initial matrix cracking followed by progressive failure of the fibers. TM was determined by the slope of the initial portion of stress–strain curve. A significant improvement in the TS and TM of PLLA was obtained by the reinforcement with untreated WJF; in warp direction their respective values were increased to 102.5% and 211.1%. In weft direction, TS and TM were also found to be 77.5% and 105.6% higher than unreinforced PLLA samples, respectively. The strains at maximum tensile stress for untreated WJF/PLLA composite samples were found to be 3.8% in warp direction and 4.1% in weft direction. It has been noticed that TS, TM and elongation properties of untreated WJF/PLLA composite are higher in warp direction than weft direction.

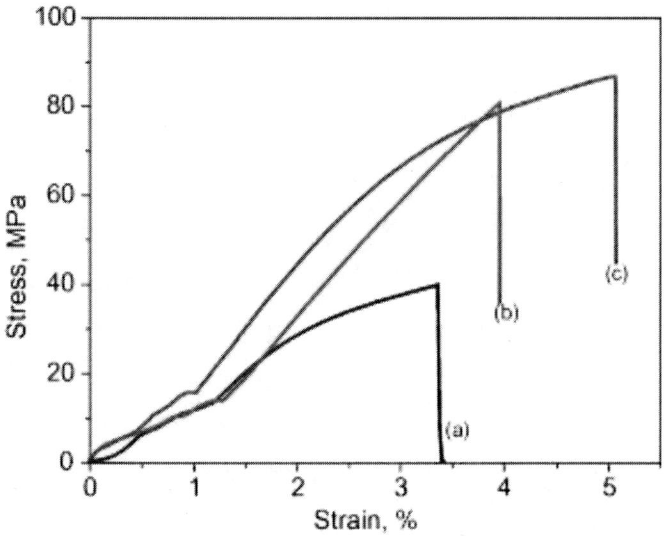

Figure 3. Tensile stress–strain diagrams of (a) PLLA, (b) untreated WJF/PLLA composite in warp direction and (c) treated WJF/PLLA composite in warp direction.

When tensile stress is applied on woven fabric composite, it induces transverse loads at fiber overlap sections of crimped fibers which attempt to straighten. This reduces the translation of fiber strength to fabric strength and decreases long-term fatigue and creep rupture performance (McDaniels et al., 2009). However, these kinds of crimp effects depend on fabric density of composite. The WJF contains more number of fibers in warp direction than in weft direction. Due to its higher yarn at warp direction composite having lower density might be the reason why its performance shows contrast with weft direction of composite. The orientation of yarns in fabric also resulted in the difference of tensile strain value. In a single layer warp woven composites, the longer yarns can extend more during the tensile test. For this reason, tensile stain is found greater in warp direction.

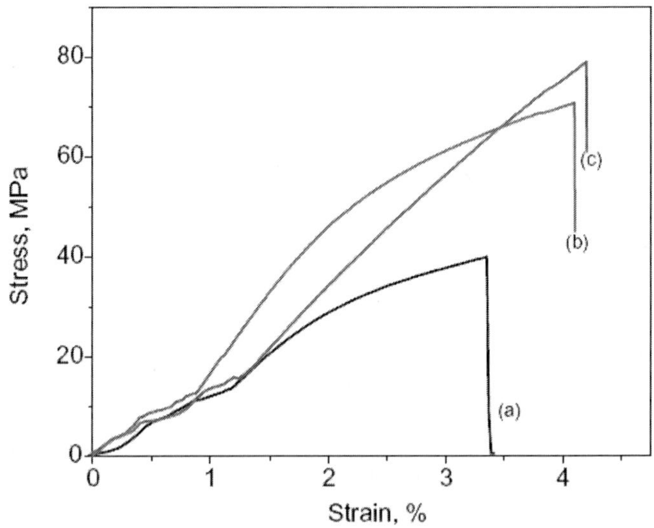

Figure 4. Tensile stress–strain diagrams of (a) PLLA, (b) untreated WJF/PLLA composite in weft direction and (c) treated WJF/PLLA composite in weft direction.

The flexural modulus is a measure of the resistance to deformation of the composite in bending. The flexural behavior of jute fabric composites is presented in Figure 5 and Figure 6. Under the flexural loading, the surfaces of the specimen are subjected to greater strains than the sample center. Hence, FS and FM are controlled by the strength of the extreme layers of reinforcement (Munikenche et al., 1999). Composites faced compressive and tensile fracture during flexural test. Surface layer of composites faced compressive fracture while the tensile mode on the bottom layer. The average FS and FM of untreated WJF/PLLA composite in warp direction were found to be 95.2% and 42.4% higher than the unreinforced PLLA. As like tensile properties, flexural behavior of WJF/PLLA ric composite is found higher in case of warp direction than weft direction. Short yarns in weft direction of composite limit their tensile stress dispersion. Therefore, as the tensile stress tries to propagate upwards, delamination failure occurred thus reducing its flexural strength (Wong et al., 2010).

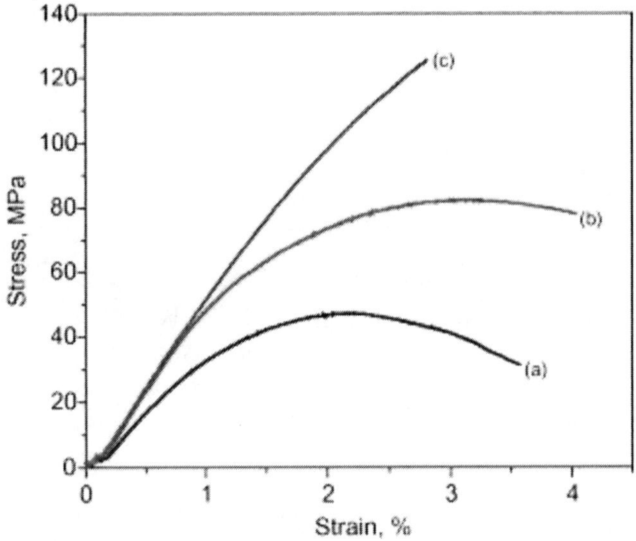

Figure 5. Flexural stress–strain diagrams of (a) PLLA, (b) untreated WJF/PLLA composite in warp direction and (c) treated WJF/PLLA composite in warp direction.

The effect of WJF on impact strength of PLLA is presented in Table 1. The average impact strength of untreated WJF/PLLA composite in warp direction (16.4 kJ/m^2) is 85.9% higher than the impact strength of unreinforced PLLA. This indicates that woven fabric plays an important role in the impact resistance of the composite as they interact with the crack formation in the matrix and act as stress-transferring medium. Containing higher aligned fibers at warp direction, the composite produced higher resistance to impact stress.

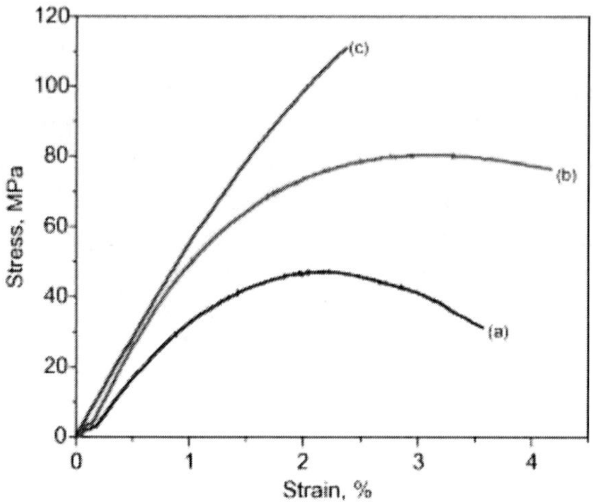

Figure 6. Flexural stress–strain diagrams of (a) PLLA, (b) untreated WJF/PLLA composite in weft direction and (c) treated WJF/PLLA composite in weft direction.

It is also seen that the mechanical properties of WJF/PLLA composites are higher than those of NWJF/PLLA composites (Table 1). In woven fabric, fiber yarns of warp direction crossover under the fiber yarns of weft direction to create an interlocking structure. Under tensile loading, these crimped fibers tend to straighten out, which create high stresses in the matrix. As a result, the strength of the woven fabric composite is greater than the strength in NWJF/PLLA composite.

Figure 3, Figure 4, Figure 5 and Figure 6 and Table 1 show that benzoylated woven jute reinforced PLLA composites possessed higher TS, TM, FS, FM and IS compared with those of untreated counterparts. It indicates that the composites possess better interfacial bonding due to fiber surface treatments. Untreated woven jute reinforced PLLA, which had poor interfacial bonding between the fabric and the matrix, showed a lower tensile strength due to the fact that a lower load is needed to break and pull-out the fiber from the matrix (Fig. 7). However, the poor interfacial bonding property between the fabric and the matrix made it easier for the fiber to debond from the matrix. The crack was somewhat blunt and the composites showed brittle properties. For treated woven jute reinforced PLLA composites, due to the improved interfacial bonding properties, the pull-out fiber from the matrix is rare and a good fracture resistance property was achieved.

Table 1. Mechanical properties of jute fabrics/PLLA composites.

Table 1 Mechanical properties of jute fabrics/PLLA composites.

Sample	Tensile strength (MPa)	Tensile modulus (GPa)	Strain (%)	Flexural strength (MPa)	Flexural modulus (GPa)	Impact strength (kJ/m^2)
Unreinforced PLLA	40 ± 6.3	0.36 ± 0.001	3.4	42 ± 9.7	3.02 ± 0.8	8.82 ± 0.9
NWJF/PLLA composite	55 ± 11.5	0.867 ± 0.02	6.01	67 ± 8.4	2.83 ± 1.1	12.98 ± 1.1
At warp direction						
Untreated WJF/PLLA composite	81 ± 13.5	1.12 ± 0.034	3.8	82 ± 12.0	4.3 ± 0.10	16.4 ± 1.8
Treated WJF/PLLA	87 ± 8.5	1.42 ± 0.047	5.1	121 ± 13.4	5.3 ± 0.10	18.1 ± 2.3
At weft direction						
Untreated WJF/PLLA composite	71 ± 8.7	0.78 ± 0.063	4.1	81 ± 9.4	3.62 ± 0.08	14.3 ± 1.5
Treated WJF/PLLA composite	79.2 ± 9	0.91 ± 0.057	4.2	111 ± 8.1	4.72 ± 0.05	16.6 ± 1.8

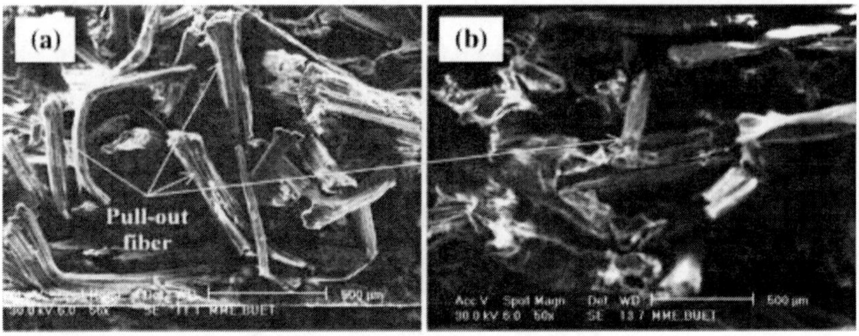

Figure 7. Scanning electron micrographs of tensile fracture surface of (a) untreated WJF/PLLA composite in warp direction and (b) treated WJF/PLLA composite in warp direction.

4. CONCLUSION

In the present research effort, mechanical tests were conducted on woven and non-woven jute fabric reinforced PLLA based composites. Based on the experimental results, the following conclusions can be drawn:

1) Woven structure exhibited excellent mechanical behavior under tensile, flexural, and impact loadings compared to non-woven composite.

2) Tensile, flexural, and impact strengths of WJF/PLLA composite were found higher at warp direction than weft direction.

3) Chemically modified WJF/PLLA composite offered better strength and modulus than untreated WJF/PLLA composites.

Finally, it can be concluded that PLLA based woven jute fabric composites might be a good alternate of synthetic fiber composites and are suitable for high load bearing applications. Further investigations have to be carried out to determine the thermal and biodegradability of these composites.

ACKNOWLEDGEMENTS

The authors are grateful to the Director, MME department, BUET, Bangladesh for his cooperation to measure SEM, which are used in this investigation.

REFERENCES

1. Ahmed, K.S., Vijayarangan, S., 2007. Experimental characterization of woven jute-fabric-reinforced isothalic polyester composites. J. Appl. Polym. Sci. 104 (4), 2650–2662.

2. Baghaei, B., Skrifvars, M., Berglin, L., 2013. Manufacture and characterization of thermoplastics composites made from PLA/ hemp co-wrapped hybride yarn prepregs. Composites Part A 50 (1), 93–101.

3. Bax, B., Mussing, J., 2008. Impact and tensile properties of PLA/ cordenka and PLA/flax composites. Compos. Sci. Technol. 68 (7– 8), 1601–1607.

4. Ganster, J., Fink, H.P., Pinnow, M., 2006. High-tenacity man-made cellulose fibre reinforced thermoplastics–injection molding compounds with polypropylene and alternative matrices. Composites Part A 37 (10), 1796–1804.

5. Graupner, N., Herrmann, A.S., Mussig, J., 2009. Natural and manmade cellulose fiber–reinforced poly(lactic acid) (PLA) composites: an overview about mechanical characteristics and application areas. Composites Part A 40 (6–7), 810–821.

6. Jekabsons, N., Bystrom, N., 2002. On the effect of stacked fabric layers on the stiffness of a woven composite. Composites Part B 33 (8), 619–629.

7. Khalil, H.P.S.A., Jawaid, M., Bakar, A.A., 2011. Woven hybrid composites: water absorption and thickness swelling behavior. BioResources 6 (2), 1043–1052.

8. Khan, G.M.A., Terano, M., Alam, M.S., 2013. Synthesis and characterization of high molecular weight poly(L-lactic acid) using

stannous octoate/maleic anhydride binary catalyst system. J. Polym. Mater. 30 (4), 397–410.

9. Lai, W.L., Mariatti, M., 2008. The properties of woven betel palm (areca catechu) reinforced polyester composites. J. Reinf. Plast. Compos. 27 (9), 925–935.

10. Lomov, S.V., Willems, A., Verpoes, I., Zhu, Y., Barburski, M., Stoilova, T., 2006. Picture frame test of woven composite reinforcements with a full-field strain registration. Text. Res. J. 76 (3), 243–252.

11. Lomov, S.V., Bogdanovich, A.E., Ivanov, D.S., Mungalov, D., Verpoest, I., Karahan, M., 2009. Damage progression in 2D and Non-crimp 3D woven composites. In: Proceedings of Composites, 2nd ECCOMAS Thematic Conference on the Mechanical Response of Composites, Imperial College, London.

12. McDaniels, K., Downs, R.J., Meldner, H., Beach, C., Adams, C., 2009. High strength-to-weight ratio non-woven technical fabrics for aerospace applications. Cubic Tech Corp., 1–9.

13. Munikenche, G.T., Naidu, A.C.B., Chhaya, R., 1999. Some mechanical properties of untreated jute fabric-reinforced polyester composite. Composites Part A 30 (3), 277–284.

14. Oksman, K., Skrifvars, M., Selin, J.F., 2003. Natural fibres as reinforcement in poly lactic acid (PLA) composites. Compos. Sci. Technol. 63 (9), 1317–1324.

15. Petersson, L., Kvien, I., Oksman, K., 2007. Structure and thermal properties of poly(lacticacid)/cellulose whiskers nanocomposite materials. Compos. Sci. Technol. 67 (10–11), 2535–2544.

16. Pothen, L.A., Thomas, S., Groeninckx, G., 2006. The role of fiber/ matrix interactions on the dynamic mechanical properties of chemically modified banana fiber/polyester composites. Composites Part A 37 (9), 1260–1269.

17. Sapuan, S.M., Leenie, A., Harimi, M., Beng, Y.K., 2006. Mechanical properties of woven banana fiber reinforced epoxy composites. Mater. Des. 27 (8), 689–693.

18. Sawpan, M.A., Pickering, K.L., Fernyhough, A., 2007. Hemp fibre reinforced poly(lacticacid) composites. Adv. Mater. Res. 29–30 (4), 337–340.

19. Shukor, F., Hassan, A., Islam, M.S., Mokhtar, M., Hasan, M., 2014. Effect of ammonium polyphosphate on flame retardancy, thermal stability and mechanical properties of alkalitreated kenaf fiber filled PLA biocomposites. Mater. Des. 54 (1), 425–429.

20. Wong, K.J., Nirmal, U., Lim, B.K., 2010. Impact behavior of short and continuous fiber-reinforced polyester composites. J. Reinf. Plast. Compos. 29 (23), 3463–3474.

Chapter 9

Sensory and Physiological Issue

Laurence Schacher[1], Sourour Bensaid[2], Selsabil El-Ghezal Jeguirim[1] and Dominique Adolphe[1]

[1] Laboratoire de Physique et Mécanique Textiles, University of Haute Alsace, France

[2] Ecole Nationale d'Ingénieurs de Monastir, Unité de Recherche Textile de Ksar Hellal, University of Monastir, Tunisia

1. INTRODUCTION

1.1. Economical, Social, Environmental, and Industrial Contexts

A significant added value of textiles is sensory in nature. During garment handling in a shop, the customer engages a selection process that involves touching and trying on clothing. A multifaceted multi-sensory, emotional and psychological experience occurs. A decision is made and motivations are based on anticipated reality of preference, personality, and emotion.

The Integration of sensory feeling in product development was not only focused in textile industry. The use of sensory analysis methods started in the sixties for food products and has been extended to many non-food products including personal care, pharmaceutical and paper products. New design processes integrating the designers' constraints and the consumers' expectations for better qualities perception have therefore been proposed in other industries. Sensory analysis methods have been widely used as market research tools in recent times and significant examples of application can be found in automotive industry. The latter has recently performed several attempts to understand and characterise the sensory preferences of end-users and to translate these features into technical specifications for new products development (Giboreau et al., 2001). However, this trend may be more prominent for textile industry as many factors can be given for conferring "character" to a material observed through handling. Micro fibres, silk-like and peach-like, cool or soft touches have been successfully developed in the past and new and exciting textile products. Hence, finishing treatments are still studied and launched on the market for that purpose. This phenomenon has been largely increased nowadays by the new textile industry developments in terms of globalization and new virtual-environment applications demand for variety and personalization. The main objective is to tailor products to the preferences of each consumer (Nakano, 1994), (Okamoto, 1991).

2. SCIENTIFIC CONTEXT – STATE OF THE ART

Recently, industrialists have moved away from usability-based approaches and towards different ones to defining user requirements. This strategy provides a framework for considering the sensory, hedonic and practical user's requirements within the product design or product evaluation process. Therefore, considering the benefits that a product should bring to its users, the next step is to determine the design characteristics through which the product can deliver these benefits. It can be considered to referring to*kansei engineering* which has been pioneered by Nagamachi (Nagamachi, 1995) in Japan since the 1970s. *Kansei* is a Japanese term for consumers' psychological impressions and feelings about a product. This approach encompasses physical, physiological, and psychological point of views.

2.1. Physical Point Of View

More than eighty years have passed since the earliest efforts of Peirce (Peirce, 1930) in the textile field to evaluate fabric hand thanks to physical measurement data. Several studies have then been undertaken to use instruments to measure fabric hand, notably Kawabata's method of the

Japanese Hand Evaluation and Standardization Committee (HESC) in the 1970s (HSEC, 1980; Kawabata, 1988), and a number of mechanical devices, including KES (Kawabata Evaluation System), FAST (Fabric Assurance by Simple Testing), and UST (Universal Surface Tester), have been developed to objectively characterize fabric hand quality (Pan & Yen, 1992; Pac et al., 2001; Fontaine et al., 2005;Issa et al., 2008; Maâtoug et al., 2009). Along the instrumental tests, all fabrics are subjected to deformations similar to those applied by the hand experts, using the same modes and rates. These tools are usually fast, repeatable and well understood; however, they may not represent the textile good in use as the measured mechanical parameters cannot directly reflect human sensation in a precise way. Indeed, fabric perception, which can be defined as the human sensory response towards fabric, obviously involves not only physical and mechanical factors, but also physiological, perceptional and social factors.

2.2. Physiological Point Of View

Physiological approach is focused on the body - information derived from the sensory organs. This information covers tactile, visual and olfactory properties. Tactile pleasures concern holding and handling a textile product. This might be relevant in the context of garments shopping. Olfactory pleasures concern the smell of the new product: the smell inside a new car may be a factor that effects how pleasurable it is for the owner. Hedonic characteristics of fabrics and garments are therefore to be considered by textile manufacturers.

2.3. Psycho-Physical Point Of View

The psycho-physical point of view relates to how the consumer mood is affected by interaction with a product. It might be expected, for example, that a new garment would provide a higher level of psychological pleasure than the old casual one. In the context of textile products it would relate to, for example, the aesthetics of a product and the values that a product embodies.

3. METHODS TO ASSESS HUMAN PERCEPTIONS: SENSORY ANALYSIS

In the 50s, sensory analysis emerged, first by the development of methods used by food industry and in 1974 in USA; a complete methodology for descriptive analysis was proposed (Stone et al., 1974). The sensory analysis has shown a promising tool for taste and smell in food industry and has been applied with success in cosmetic industries such as Nestle, L'Oreal, Dior, (Young et al., 2005; Stone & Sidel, 2007). Textile materials have been a subject of interest concerning sensory analysis. The first attempt is reported by Binns (Binns, 1926). Since the early 80s, standard methods have been developed and published. They are customization of methods established in food science (Depledt, 1998); Barthelemy et al., 1990; Meilgaard et al., 1991). A methodology for sensory analysis of tactile feeling of textile fabrics was developed in France and the results reported a creation of tactile sensory profile (Depledt, 1998; Cardello et al., 2003; Philippe et al., 2003; Chollakup et al., 2004 a; 2004 b; Bensaid et al, 2006). Automotive industry has also applied sensory methods for their own products. Sensotact® reference frame is a commercial example of an attempt to formalize and calibrate descriptions of tactile perceptual dimensions. It was developed by French Renault Automotive Company (Sensotact, 2008). Italian Fiat Company (Bandini et al., 1997) has also shown some relevant sensory design engineering examples for their products.

3.1. Definition Of Sensory Analysis

The basic assumption of sensory evaluation is the ability to perform objective measurements of sensations using a panel of people as an instrument. The sensory analysis is defined as « the examination of the organoleptical properties of a product using the human senses » (ISO 5492, 1992). Fortin and Durand (Fortin & Durand, 2004) give the following assertion « The sensory analysis can be defined as the study of the human response to a stimulus (...) The sensory analysis qualifies and quantifies the felt perceptions of persons called judges or panelists when they evaluate products or materials inducing our reactions senses. These methods could be applied to food, perfumes, cosmetics, textiles, automotives... Based on these definitions, it can be assumed that the sensory analysis of the products consists in the description or their evaluation through words called descriptors or attributes linked to each of our senses (sight, hearing, taste, smell, touch).

3.2. Measurement Principle

The evaluation starts with the contact between the body and the environment and the interaction of the product with the judge. It is the "sensorial stimuli", which will be analyzed in different manners regarding the personal history of the "judge" and will be linked with the relationship existing between him or her and the evaluated object. The analysis of the "sensorial stimuli" is a unique function dedicated to the human being and no measurement device could reproduce it" (Fortin & Durand, 2004). In other words, the sensory analysis method uses the human senses as measurement device. In this way, it is essential to understand how the senses are working in general and the tactile sense in particular. Every sensory information is issued from a stimulus applied on the sensorial receptors by the environment. Generally, the stimulus feeling unrolls the following procedure (Rosenweig & al., 1998;Richard & Orsal, 2001).

Signal reception: every sensorial system is activated when a stimulus is generated by an external event applied on an organ. These sensory organs select stimuli, it is the filter phase.

Transduction: the human sensors convert the external event in electric signals understood by the nervous system. This step is followed by the transmission of this signal to the nervous centers.

Integration: The different information will be treated and computed by the central nervous system which will generate the response in relation to the solicitation.

The sensorial sensors are usually grouped in sensorial organs and are specialized in one particular sensorial solicitation.

A feeling is beginning when a stimulus activates a sensory organ. The sensory responses are generated after 0.1 to 0.2 second after the stimulation. First, the stimulus is coded in terms of quality and intensity. The response to a sensory stimulation could be physiological, behavioral, verbal, or psychological; taken into consideration that, at least, only the three first responses are observable (SSHA, 1998).

Generally, as a first step in fabric sensory analysis, monosense approach is preferred. Usually, tactile only or visual only examinations are performed by panelists: for tactile evaluation, tests are carried out in "blind" conditions in order to reduce biases that could be induced by seeing the fabric such as subjective preference of a special colour or material. However, human beings are equipped with multiple sensory channels through which they experience objects in the environment. There is obvious evidence that perception of information provided within one sensory modality can be greatly influenced by stimuli caused by another modality. It is true that consumer can see the fabric when touching it and tactile exploration of a textile surface is usually accompanied by visual sensory inputs and in a context of purchase. In such conditions, the consumer has the possibility to see the colour of the fabric and to know which kind of material it has been made from as well as its context of use. During tactile exploration, one can also sometimes hear the sounds made as the fingers explore the fabric. One can sometimes smell the odour of the fabrics. For all possible combination, the correlation between visual and tactile properties has mainly been studied (Lederman et al., 1981; Konyo et al., 2002, Cinel & al., 2002, Guest & Spence, 2003, Mucci et al., 2005, Bensaid & al., 2008). In these studies, the superiority of vision in the multi-modal sensational perception has been somehow demonstrated (Konyo et al., 2002). This first modality corresponds to several marketing results showing that vision is the very first data required by consumers and that the risk of not feeling, and trying on clothing before purchase may be the greatest challenge for Internet clothing sales, and is an issue that must be addressed. Some companies are developing Virtual 3-D try-on technology that might reduce the risk of ill-fitting or inappropriately styled clothing for one's body type, by providing the consumer with a view of the garment on his/her body. Inappropriate tactile or sound or smell feelings would be more complex to address with such technologies. However, Lederman et al. (Lederman, 1981; Lederman & al., 1986) have shown that the extent to which the data from one modality is preferred over the other depends on the nature of the task to be performed. Consequently, some tasks appeared best suited to vision (e.g. determining the spatial density of texture elements), and some to touch evaluation (e.g. determining the roughness of fine textures).

The touch feeling is known to be one of the most important senses. The skin which is the main organ of the tactile sensibility covers our body with an average surface of $1.7m^2$ for an adult person. It is made of two layers: a top layer called epidermis which is renewed every 20 days and a bottom layer called dermis which contains nerves and blood vessels. The touch feeling groups three main modalities: mechanical sensibility, thermal sensitivity and pain.

The mechanical sensibility: also called touch sensibility corresponds to the response to pressure, touch and vibration solicitations. This sensibility depends on numerous parameters such as shape, surface, duration and intensity of the stimulus. The mecano-receptors are the key elements of this sensibility.

Six kinds of receptors have been listed, the Merkel disks, the tactile disks, receptors of the hair follicles, the Meissner corpuscles, the Ruffini corpuscles and the Pacini corpuscles as shown in the figure 1.

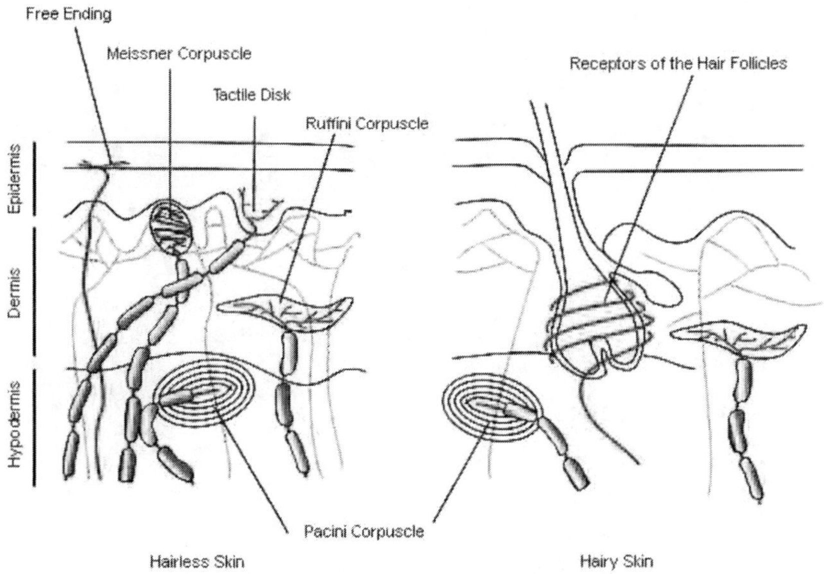

Figure 1. Scheme of the skin and location of the touch sensors (**Richard & Orsal, 2001**)

The figure 2 displays the location (a), the size and the density (b) of the four kinds of receptors in the internal face of the hand (Richard & Orsal, 2001). Every cerebral region manages a precise skin part. This repartition has been presented in a caricature manner through the “homoculus” display (figure3) where every organ is represented with a surface in relation with its touch sensors density (Richard & Orsal, 2001). Nevertheless, the tactile sensibility is not fully linked with the density of the touch sensors but it is more driven by the number of connections in the sensitive cortex of the brain. On apes, some studies have shown that the sensitive surface could be increased or decreased thanks to training.Figure 4 represents a map of the spatial discrimination for the different organs. The more sensitive parts are the fingers tips and the tongue end.

Thermal sensitivity corresponds to two opposite sensibilities, the sensitivity to warm and the sensitivity to cold. These feelings are detected by some specific points which are sensitive to hot and warm and that are scattered all over the skin with a lower density than those dedicated to mechanical stresses. Moreover, the sensitive points to cold are more numerous than those sensitive to hot.

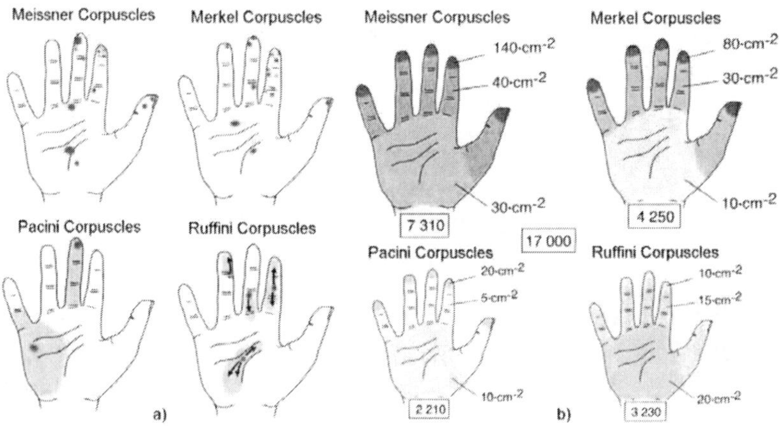

Figure 2. Touch Sensors location a) and density b) in the inside hand

Pain is a feeling generated by a high level stimulation called "novice" because it induces injuries in the organism. Different kinds of pain can be detailed: superficial pain which is coming from the skin, deep pain coming from the bones, from the muscles, from the joints or the tendons. The receptors involved in this feeling are called "nociceptors".

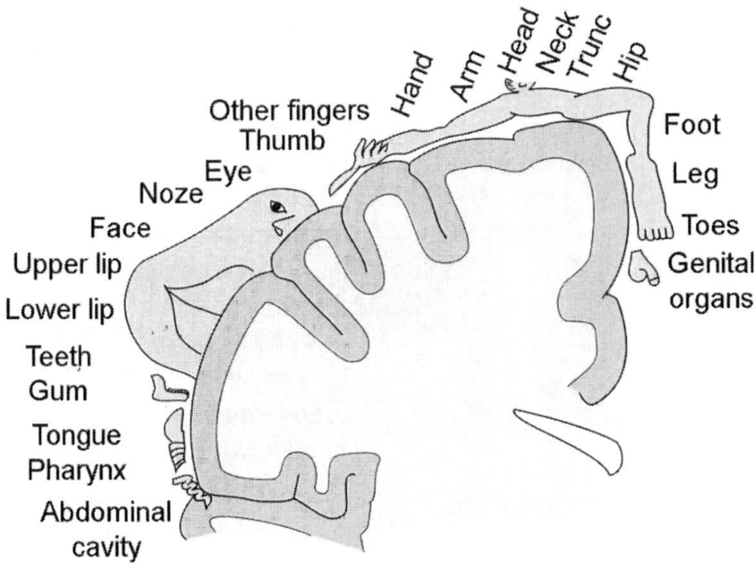

Figure 3. Homoculus caricature display

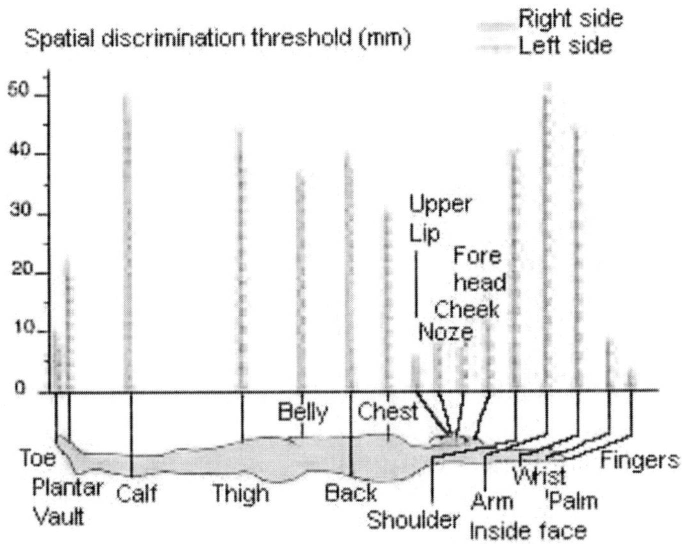

Figure 4. Spatial discrimination threshold of human organs

3.3. Sensory Analysis: Methodology

In regard to the final goal of the evaluation, comparison, quality measurement, new product characterization, etc, different kinds of tests issued of the sensory analyze set could be performed (AFNOR V09-001, 1983; AFNOR XP V 09-501, 1999; ISO 6658, 1985). One of the most frequently used, the descriptive test allows characterizing, comparing and quantifying differences between tested products. The method requires a group of trained judges (panelists), who are intensively trained to qualify and quantify their feeling and hedonic judgment in an objective way. The product characteristics are dissected through their expertise. The panelists first agree on specific and objective sensory vocabulary (ISO 11035, 1995) to describe the feeling of a product range (Nogueira & al., 2009).

Continuous training enables them to sharpen their perceptions and to perceive differences of intensity between products for every criterion. Training the panels also leads to homogeneity and reproducibility in the ratings (ISO 8586, 1993). Finally, a final grading enables to build out product sensory identity maps or profiles, or fingerprints. The obtained results are rich and powerful as they allow identifying all of a product's perception characteristics along with measures of the importance of those characteristics.

Figure 5 represents a sensory map of raw cotton woven fabrics before (non-treated) and after various commercial post-treatments (m, sd, sp, k) evaluated by 11 trained panelists. The tests were replicated twice and the order of presentation of the samples was randomized for each assessor. The displayed differences on figure 5, even the smallest one, are significant in regards to statistical analysis.

However, the panelists are not representative of the end-users and therefore are not asked to perform the hedonic studies. The complete sensory methodology process involves another group of persons: the final consumers. This group is not trained, the persons being merely questioned on their preferences (like-dislike). Both approaches lead to preference mapping (Schlich, 1995) that allows specifying preferred sensory characteristics of products for given groups of end-users. Figure 6 represents the main methods in sensory analysis.

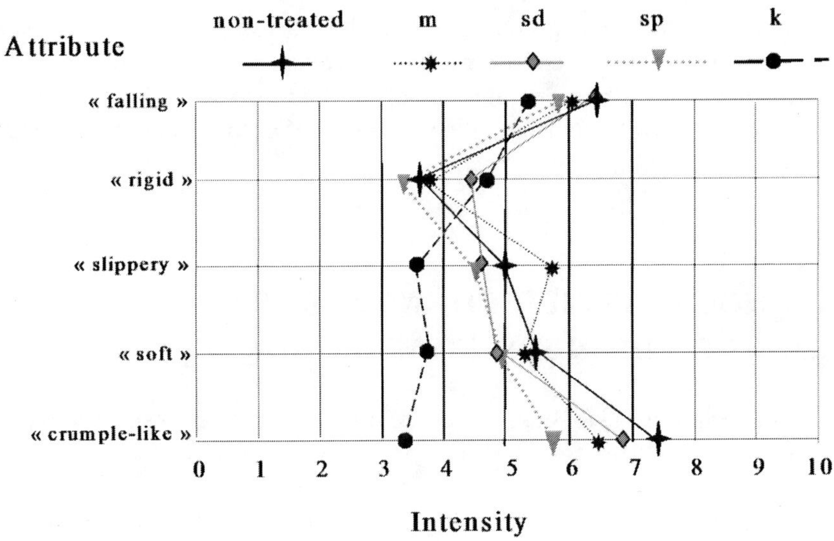

Figure 5. Example of tactile sensory profiles of textile woven fabrics

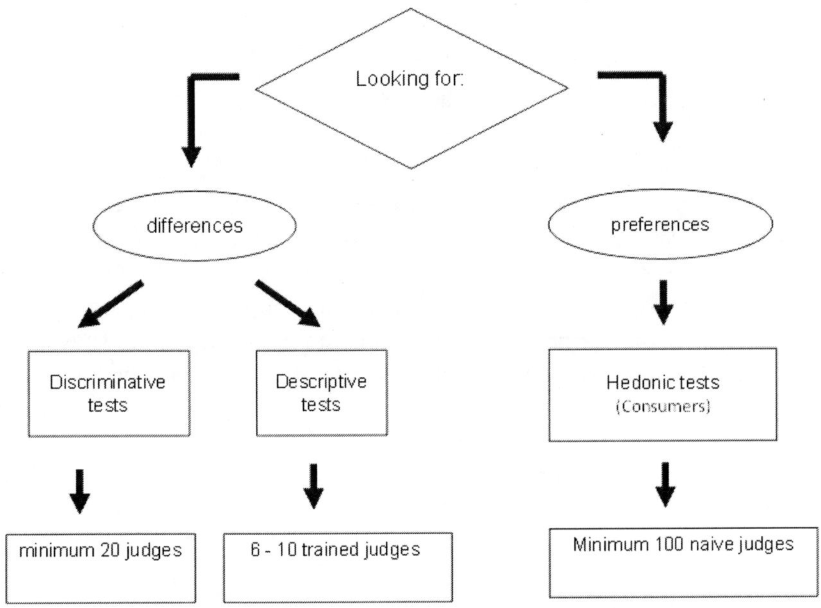

Figure 6. Methods of sensory analysis

Sensory analysis is using “human beings” as a tool but it is employing objective methods to collect their subjective sensory responses. One disadvantage of the sensory methodology is the time consuming due to the panel recruitment, training and the validations of each step of the methods using the appropriate statistical tools.

4. INNOVATING METHOD AND NUMERICAL TOOL TO SIMULATE COMPLEX SYSTEMS

Several attempts have been made to model the relationship between tactile sensory attributes of fabrics and their production parameters, or their instrumental measurements. Hence, statistics and multivariate analysis (Bishop, 1996) and methods based on intelligent techniques (neural networks, fuzzy logic...) have been used (Vassiliadis et al., 2010). The intelligent techniques have proven a very efficient tool for the fast and precise solution. Therefore, they have found increasing applications in the textile field such as properties prediction and process optimization.

4.1. Artificial Neural Network

An artificial neural network (ANN), usually called neural network (NN), is a mathematical model or computational model that tries to simulate the structure and functional aspects of biological neural networks. A neural network consists of an interconnected group of artificial neurons and processes information using a connectionist approach to computation. Neural networks are non-linear statistical data modelling tools. They can be used to model complex relationships between inputs and outputs or to find patterns in data.

There are many different ANN structures and learning algorithms available in the literature (Haykin, 2000). Among these algorithms, multilayer perceptron (MLP) has been successfully applied. A typical multi-layer neural network with a single hidden layer is shown in Fig.7. Each neuron receives a signal from the neurons of the previous layer and these signals are multiplied by separate synaptic weights. The weighted inputs are then summed up and passed through a transfer function, which converts the output to a fixed range of values. The output of the transfer function is then transmitted to the neurons of next layer. This process is continued and finally the output is produced at the output node. Predicted output is then compared with the desired output and an error signal is generated.

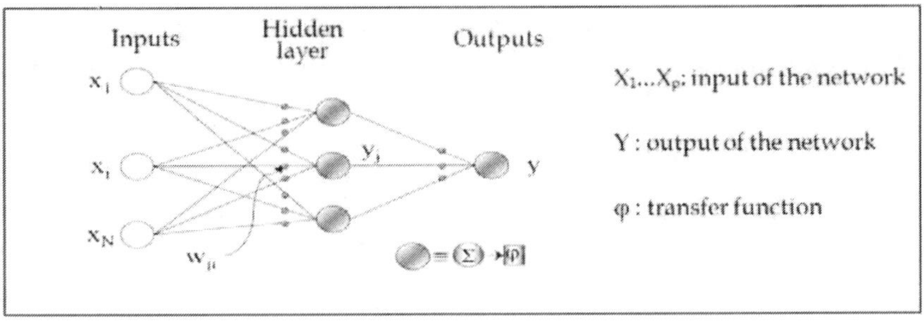

Figure 7. A multilayer artificial neural network

The error signal is then minimised in iterative steps by adjusting the synaptic weights using a suitable training algorithm (figure 7). Among the various kinds of algorithms for training neural network, the back-propagation algorithm developed by Rumelhart et al. (1986) is the most widely used. Network weights are adapted iteratively until some appropriate stopping criteria are met and the best weight vector that corresponds to the best generalization is achieved.

4.2. Fuzzy Interference System

The foundation of fuzzy logic, which is an extension of crisp logic, was first proposed by Zadeh (Zadeh, 1965). The theoretical aspects of fuzzy logic and fuzzy arithmetic have been explained in many standard textbooks (Zimmerman, 1996). In crisp logic, such as binary logic, variables are true or false, black or white, 1 or 0. In fuzzy logic, a fuzzy set contains elements with only partial membership ranging from 0 to 1 to define uncertainty of classes that do not have clearly defined boundaries. For each input and output variable of a fuzzy inference system (FIS), the fuzzy sets are created by dividing the universe of discourse into a number of sub-regions, named in linguistic terms (high, medium, low etc.). If X is the universe of discourse and its elements are denoted by x, then a fuzzy set A in X is defined as a set of ordered pairs as A={x, μA(x)|x∈X} where μA(x) is the membership function of x in A.

Once the fuzzy sets are chosen, a membership function for each set is created. A membership function is a typical curve that converts the input from 0 to 1, indicating the belongingness of the input to a fuzzy set. This step is known as "fuzzification". Membership function can have various forms, such as triangle, trapezoid, sigmoid and Gaussian.

The linguistic terms are then used to establish fuzzy rules. Fuzzy rules provide quantitative reasoning that relates input fuzzy sets with output fuzzy sets. A fuzzy rule base consists of a number of fuzzy if-then rules. For example, in the case of two inputs and single output fuzzy system, it could be expressed as follows:

$$\text{If } x \text{ is } A_i \text{ and } y \text{ is } B_i \text{ then } z \text{ is } C_i \quad (1)$$

where x, y and z are variables representing two inputs and one output; Ai, Bi and Ci, the linguistic values of x, y and z respectively.

The rule base contains linguistic rules that are provided by experts. It is also possible to extract rules from numerical data. Once the rules have been established, the FIS can be viewed as a system that maps an input vector to an output vector.

The output of each rule is also a fuzzy set. Output fuzzy sets are then aggregated into a single fuzzy set. This step is known as "aggregation". Finally, the resulting set is resolved to a single output number by "defuzzification". These main steps involved in modeling a fuzzy system are shown in Fig.8.

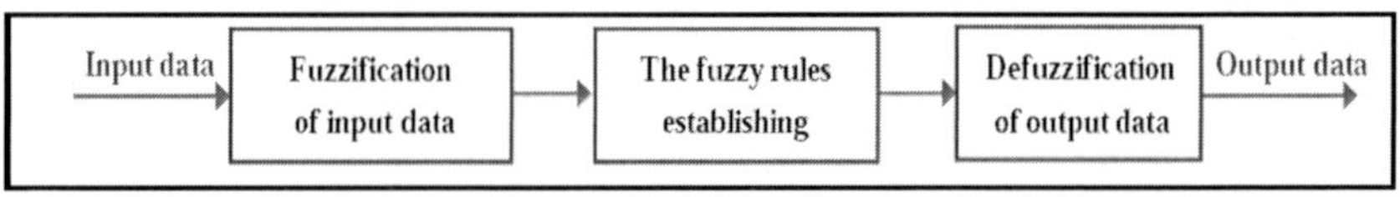

Figure 8. General depiction of fuzzy system

4.3. Modeling The Relationship Related To Sensory Evaluation

4.3.1. Modeling The Relationship Between Sensory Properties And Production Parameters

In order to face the competitive environment, textile companies are interested in designing and producing new industrial products adapted to the specific needs of consumers with a minimal number of experiments. In fact, specific fabrics tactile properties desired by consumers can be reached by adjusting the process parameters.

Several investigators have used statistics and multivariate analysis, such as multiple factor analysis MFA and principal component analysis PCA for studying the influence of finishing treatments on low stress mechanical properties and sensory attributes (El-Ghezal Jeguirim et al. 2010 a; 2010b; 2011). Although classical computing techniques are relatively efficient to interpret and analyze the relationship between sensory properties and production parameters, some limitation related to the non-linear relations in sensory domain has been reported (Zeng et al., 2008).

New methods based on intelligent techniques (fuzzy logic, neural networks...) are used to treat a great number of textile applications (Dubois & Prade, 1997; Kwak et al., 2000; Jain et al., 2004; Wong et al., 2006; Ertugrul & Ucar, 2000; Vassiliadis et al., 2010). Zeng et al. have used the fuzzy logic technique for modeling the relationship between the production parameters and the physical features of fabrics (Zeng et al., 2004). The instrumental features have been measured on Kawabata Evaluation System. In order to reduce the inputs number, a small number of relevant physical features have been selected using human knowledge on fabric production and fabrics properties. In the modeling procedure, the fuzzy rules have been extracted from measured numerical data. These extracted rules have been validated and adjusted by human knowledge on production processes. In this way, the two information sources (human knowledge and measured data) are both taken into account in the fuzzy rules of this model.

El-Ghezal Jeguirim et al. (El-Ghezal Jeguirim et al., 2009) have developed neural network and fuzzy logic based models to predict the sensory attributes, evaluated by a trained panel, of knitted fabrics from the structure and process parameters. In their further work, the intelligent techniques have been used for modeling the relationship between the instrumental properties measured by Kawabata Evaluation System and the finishing parameters of knitted fabrics (El-Ghezal Jeguirim et al., 2011). The prediction performance of these models was considerably lower than the mean variations of experimental values. These results showed the intelligent techniques ability to model the relationship between manufacturing parameters and instrumental or sensory tactile properties. The fuzzy or neural models provide contribution in industrial products engineering, with minimal number of experiments and short cycles of product design. The prediction performances of neural and fuzzy models were also similar. However, the 'black box' problem associated with neural networks can hinder the widespread adoption of this method. In fact, the fuzzy techniques have two advantages over the neural ones. In fuzzy models, the linguistic rules can be interpreted and the linguistic sensory attributes can be integrated. Thus, it is possible to observe how the fuzzy model performs its computations.

Moreover, better results can be obtained when these techniques are used in combination. In fact, hybrid models based on neuro-fuzzy methodologies combine the self learning ability of neural networks and the human-like reasoning style of fuzzy systems. Hence, neuro-fuzzy models can provide enormous scopes to link sensory attributes or mechanical properties with processes parameters of fabrics.

4.3.2. Modeling The Relationship Between Sensory Properties and Instrumental Measurements

The sensory evaluation method has been adapted to textile products to characterize consumer preference for textile products. However, sensory evaluation is time-consuming and expensive. Therefore, reliable and practical instrumental methods are needed to accurately predict sensory tactile attributes, at least in the product development and quality control stages. Several instrumental methods (Kawabata, 1975; 1980; Kawabata et al., 1982; Tester and De Boos, 1990; Pan & al., 1993, Pan & Yen., 1992) have been developed for measuring the tactile feeling of fabrics according to their physical mechanical, thermal and surface properties. Hence, modeling the relationship between instrumental measurements and sensory attributes becomes a key problem in tactile quality assessment.

Wide ranges of statistical or multivariate analysis have been proposed for characterizing this relationship and selecting the relevant instrumental and sensory properties. In particular, Weber-Fechner's law (Matsuo et al., 1971; Bishop, 1996; Rombaldoni et al., 2010), Stevens's power law (Bishop, 1996; Elder et al., 1984; Rombaldoni et al., 2010), and PCA (Mackay el al., 1999; El-Ghezal Jeguirim et al., 2010 b) have been usually applied.

Matsuo et al. (Matsuo et al., 1971) used the Weber-Fechner's law to translate instrumental measurements of a wide of fabric mechanical properties into corresponding hand parameters. Although the method has the virtue of independence of fabric type, further work would be needed to check the validity of the Weber-Fechner's law for modeling the complex sensory-instrumental relationship.

Elder et al. (Elder et al., 1984) used Stevens's power law to examine the relationships between subjectively measured softness and a compression and also between subjective stiffness and a flexural rigidity. Excellent correlation was found, correlation coefficients for the Stevens's law being about 0.97. On the evidence of the results of Elder et al., Stevens's law appears to be an excellent model. Although the relationship breaks down in some cases, this fact is probably because the subjective evaluation attribute cannot be adequately represented by a single instrumental parameter. This problem may be overcome by relating each sensory score to the sum of the different contributions made by a number of instrumentally measured properties that are relevant to well-defined fabric types or end uses. Rombaldoni et al. (Rombaldoni et al., 2010) investigated the possibility of predicting the human psychophysical perception of crispness and coolness hand of men's suit woven fabrics made from animal fibers (wool, mohair, cashmere and alpaca) from measurable low-stress mechanical and thermal parameters. In particular, the parameters chosen were weight per unit area, thickness at 9.81 kPa, surface thickness, bending rigidity, extensibility at 98.1 N/m, shear rigidity, formability and thermal absorptivity. The sensory-instrumental relationship was explored using the Stevens's power law. The correlation results were also compared by the predictive power of other mathematical models: a linear function and the Weber–Fechner law. The obtained results showed that the Weber–Fechner-law-based model was the best to predict the sensory hand value.

Mackay el al. (Mackay et al., 1999) used Principal Component Analysis (PCA) to study relationships between sensory and instrumental measurements of the effect of washing processes on 1x1 rib knitwear fabrics. El-Ghezal Jeguirim et al. (El-Ghezal Jeguirim et al., 2010 b) investigated the relationship between instrumental data and sensory attributes, assessed by a trained panel by using PCA. The obtained results have shown that the compression resilience, the geometrical and frictional roughness are significantly correlated with the following sensory parameters thick, heavy, soft, elastic and crumple-like attributes.

The intelligent techniques, including fuzzy logic and neural networks are also used for modeling the relationship between instrumental measurements and sensory properties. Hui et al. have developed a neural network to predict the consumers sensory data from fabric properties (Hui et al., 2004). The predicted results are highly correlated to the targets in the fabrics made by five materials: cotton, wool, polyester, nylon, and acetate. In the validation, the results show that the proposed network also predicts the hand of linen and silk. Another work for predicting fabric hand from physical measures has been done by combining fuzzy logic and neural networks (Park et al., 2000). The obtained results revealed that the neural networks and fuzzy logic provide an alternative approach for predicting sensory properties from instrumental measurements of fabrics.

5. APPLICATIONS STUDY CASES FOR COTTON/WOOLEN WOVEN FABRICS WITH DIFFERENCE PATTERN

The main applications of sensory analysis in textile industry are focused on the analysis of the factors affecting fabric hand. Many studies have been conducted to identify the effect of construction factors, such as pattern, yarn properties, fabric materials, or finishing treatments on the fabric tactile quality by using the sensory analysis (Bensaid & al., 2006; El-Ghezal Jeguirim et al., 2010 a; 2010 b).

In order to provide reliable descriptions and evaluation of fabric hand, rigorous procedures have to be implemented i.e. exploratory procedures, samples presentation and data analysis. In this section, a detailed description of the followed procedure is presented before the discussion of some studies' results.

Description of the followed experimental protocol

Sensory analysis method has been developed in the group since 1999. It requires a group of trained persons. In these cases, the sensory panel comprises nine adults between 20 and 50 years old. These persons, called assessors, have been trained according to a previously established methodology (Nicod, 1990; NF-ISO, 1993). The tests are conducted in an environmentally controlled room (20º C and 65% Relative Humidity) (NF EN 20139, 1992). The assessors evaluate the products twice, in particular conditions, in order to avoid some bias due to other senses. The presentation of the fabrics is randomised and the evaluation is done without seeing the fabric using specially designed booth.

Sensory rating is done using a set of 15 individual sensory attributes (Table 1) to build profile consisting of the descriptive, quantitative and objective analysis of the fabric. These attributes have been consensually selected by the assessors and have been used for different types of fabrics (NF-ISO 5492, 1992; NF-ISO 11035, 1995). Quotation is performed on a non-structured scale 0-10.

Before every new product category evaluation, the assessors are retrained for ten sessions. This step allows them to become familiar with the procedure of evaluation and to remember the right meaning and extremes of each attribute.

The pertinence of the attribute is checked later using statistical tools.

In this chapter the effect on the fabric hand of cotton fabrics of three series of parameters will presented: effect of the weaving patterns, effect of the yarn count, and effect of finishing treatments.

5.1. Effect Of The Weaving Patterns On The Fabric Hand Of Cotton Fabrics

Materials

In order to study the effect of weaving pattern, nine fabrics have been selected. The samples have been woven with nine classical weft effect patterns on a Jacquard loom. The patterns include a plain weave, 3-twill weave, 4-twill weave (Z direction), waved twill weave (horizontal effect), 4-satin weave, 5-satin weave, 6-satin weave, 12-satin weave and crêpe weave; the yarns are 100% cotton, 14 Tex for the warp and 25 Tex for the weft. In order to highlight the influence of the pattern on the tactile feeling, we tested fabrics having close characteristics. Hence, the saturation index in weft direction was kept equal to 52%. This index is defined as following:

$$I_r = \frac{100 \text{ x D (yarn count within the fabric)}}{\text{D}_{max} \text{ (maximum yarn count)}} \quad (2)$$

Table 1. List of the 15 attributes used for the sensory evaluation

Bipolar attributes	Surface attributes	Handle attributes
cold-warm thin-thick light-heavy supple-rigid	pilous soft granulous sticky grooved greasy slippery	falling responsive crumple-like elastic

Microphotographs of the fabrics weaved in different patterns are presented in Figure 9. Their characteristics are presented in Table 2.

Table 2. Characteristics of the tested fabrics

Pattern	Weft count (tex)	Warp count (tex)	Area density g/m²	Saturation index (%)
Plain	25	14	124.8	52
3-Twill	25	14	140.0	52
4-Twill	25	14	140.8	52
4-Satin	25	14	141.8	52
5-Satin	25	14	141.6	52
6-Satin	25	14	147.0	52
12-Satin	25	14	147.2	52
Waved twill	25	14	150.2	52
Crêpe	25	14	135.6	52

Results and discussion

Statistical methods of data analysis have been applied. The ANOVA 2-way test (5%), applied on product across assessor variables, outlined that only 4 attributes are not affected by the fabric pattern (warm, sticky, greasy and elastic), the other 11 attributes are significantly affected (Table 3). Four attributes are considered as non pertinent for this study: cold–warm, sticky, greasy and elastic. The three first are material dependant and the same weft and warp material (100% cotton) was adopted whatever the pattern is. The last one, elastic, could be dependant of the pattern, but in woven fabrics case, the elastic behavior is relatively low in comparison to the knitted fabrics. Hence, the marks given by the panellists are around 0 with a non significant difference between the different samples.

Figure 9. Microphotographs of the nine different patterns

According to the column PdxAs, it appears that soft, granulous, pilous, grooved and crumple like have been marked in a different manner depending of the judges.

The sensory profile is the simplest way to visualize the results. Each horizontal line represents the scale of one attribute, on which the mean score is plotted. All marks related to the same product are joined by a line. Figure 10 shows the profiles of the nine weaving patterns.

Table 3. ANOVA 2-way test (5%) for all the attributes and all the fabrics

Attribute	Product	Assessor	PdxAs
Warm	NS	×	NS
Falling	×	×	NS
Thick	×	×	NS
Heavy	×	×	NS
Rigid	×	×	NS
Sticky	NS	×	NS
Slippery	×	×	NS
Soft	×	×	×
Granulous	×	×	×
Greasy	NS	×	NS
Pilous	×	×	×
Grooved	×	×	×
Elastic	NS	×	NS
Responsive	×	×	NS
Crumple-like	×	×	×

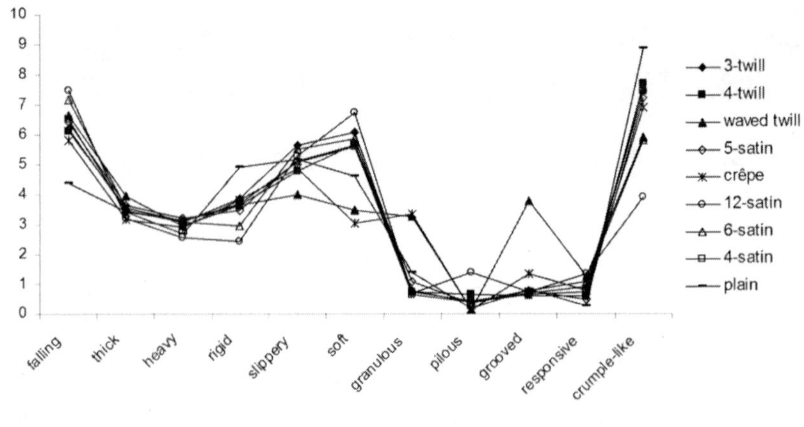

Figure 10. The sensory profiles of the nine fabrics for the relevant attributes

According to this figure, it can be observed that the waved twill, the crêpe, the plain, and the 12-satin weaves have distinguished profiles.

The Friedman test (5%) ranges the products into groups. The obtained results show that the nine fabrics are close, and they belong to the same group for cold–warm, light-heavy, sticky, slippery, greasy, and elastic attributes, since the differences between fabrics are not significant. However, the fabrics are divided into several groups for falling, thin-thick, supple-rigid, soft, granulous, pilous, grooved, responsive and crumple-like attributes.

The Principle Component Analysis (PCA) is one of the most frequently used methods for the analysis of data collected from sensory tests. It is applied on the mean score of the panel across replication to analyze the pertinence of the attributes and to obtain graphical displays of the multivariate data simplifying subsequent analysis and highlighting similarities and differences between the woven fabrics. Sensory attributes were abstracted into two sensory independent factors, which explain respectively 46% and 28% of the total variance. These groups are carried to the map of products, in order to see the different correlations between fabrics and attributes. The results are presented in Figure 11.

In this figure, it can be observed that:

- the plain weave is the most rigid and crumple-like and the least falling and responsive.
- the waved twill is the most granulous and grooved and the least soft and slippery.
- the 12-satin is the most pilous, responsive, falling, soft, and slippery.
- the crêpe has a rigid feeling; it is also less pilous, soft, and responsive than most of the other fabrics.

Conclusion

Based on the obtained results, and as predicted, it is seen that the pattern strongly influences tactile feeling. Several attributes have been affected: soft, slippery, grooved, granulous, pilous, rigid, and falling. These results are in accordance with textile professionals' expectations.

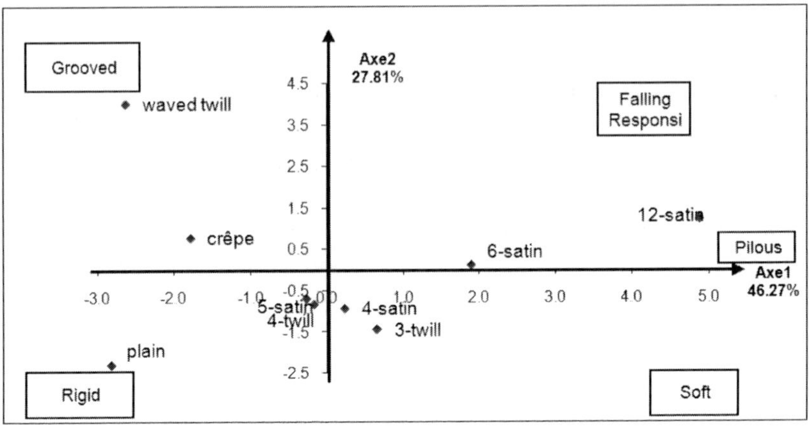

Figure 11. Principle Component Analysis, map of products

Plain weave, waved twill, crêpe and 12-satin have very distinguished tactile profiles as compared to the other fabrics.

Knowing the correlations that may exist between fabric pattern and tactile properties, manufacturers would be able to design specific touch by the weaving process instead of using finishing treatments. This may be interesting in order to develop an environmental friendly process and avoid the use of chemical products.

5.2. Effect Of The Yarn Count On The Fabric Hand Of Cotton Woven Fabrics

All woven fabrics are made by yarns. It is therefore interesting to study the effect of yarn properties on fabric hand. In this paragraph, on yarn property, the yarn count is studied. The impact of this factor on fabric sensory properties is underlined.

Materials

The study is carried out on 4 fabrics having different weft counts: 25 Tex, 50 Tex, 71 Tex and 100 Tex. The other parameters are unchanged: 100% cotton, Warp count (14 Tex) and Index of saturation (52%).

The experiment is applied on nine different patterns. Only the results of the plain weave are presented in this paragraph.

Results and discussion

The ANOVA 2-way test (5%) revealed that 9 attributes are significantly affected. These attributes are: thin-thick, light-heavy, supple-rigid, soft, granulous, grooved, falling, responsive and elastic.

The sensory profiles are presented in Figure 12. It can be noticed that some attributes are positively correlated to the yarn count. These attributes are: thin-thick, light-heavy, supple-rigid, granulous and responsive.

On the map of products obtained by the Principle Component Analysis (Figure 13), it can be noticed that fabrics are ranked on one principle axe (79.53%). On the left side of the axe, there are fabrics with high yarn counts and they are correlated to thick, heavy, rigid, granulous, grooved, elastic and responsive attributes. The right side contains fabrics with low yarn counts which are positively correlated to falling, thin, light, supple and soft attributes.

Those results are proven for the all other patterns: twills, satins and crêpe.

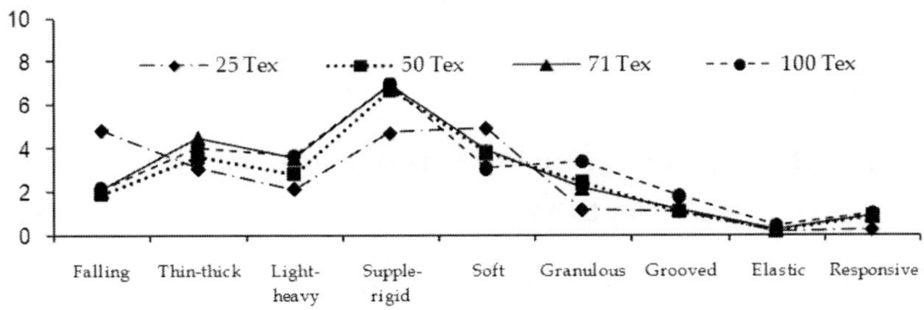

Figure 12. Sensory profiles of plain weave fabrics with different yarn counts

Conclusion

The study of the influence of yarn count on the touch quality of fabrics has been proven as very important and has shown very interesting results. Surface properties as well as full hand properties are strongly affected by the yarn count. The more the yarn count is important, the more the fabric is granulous, grooved, thick, heavy, rigid and responsive.

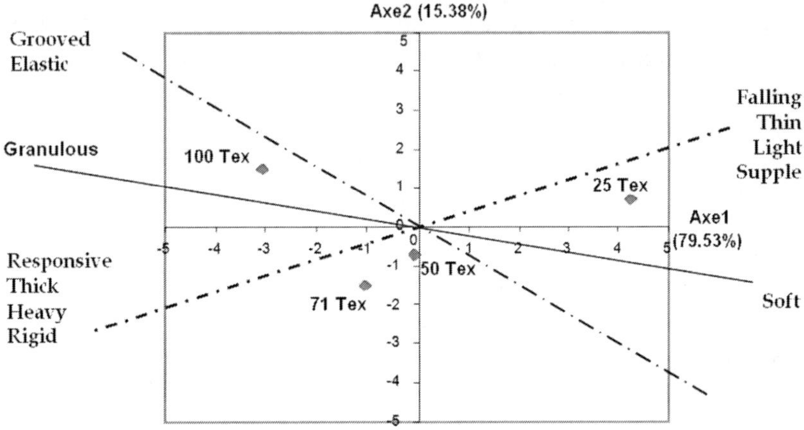

Figure 13. Map of products, yarn count effect

This may help to control and evaluate fabric tactile properties by modifying the yarn characteristics and parameters.

5.3. Effect Of Finishing Treatment On The Fabric Hand Of Cotton Woven Fabrics

In order to confer a variety of looks and effects on fabrics, there are many new finishing products and treatments proposed by chemical suppliers. This investigation was aimed by the fact that differences between fabric treatments technologies could be distinguished more evidently than it was done before thanks to sensory evaluation methods.

Testing methods and materials

The tests are carried out on 100% cotton plain weave fabric, 24 yarns/cm weft and warp, 160 g/m^2, scoured and bleached. Two finishing products were studied: the crease-resistant finishing Knittex "K" and the softener macro silicone Ultratex® "Ul".

Knittex® FEL: a nonionic crosslinking resin based on a modified dimethyloldihydroxyethylene, allows bringing properties of anti-crease and anti-shrink to the fabric.

Ultratex® UM: cationic emulsion of functional polydimethylsiloxane, allows bringing a very soft touch to the fabric.

The products were processed using semi-industrial range and with varied concentrations of the two products (Table 4). Fabrics were tested and evaluated under controlled environmental conditions following the previously described procedure.

Table 4. Different finishing treatments

Product	Product code	Concentration (g/l)
Non treated fabric	0	0
Knittex® FEL "K"	21	20
	22	50
	4	80
Ultratex®UM "Ul"	23	5
	24	20
	17	40

Results and discussion

Seven attributes are significantly affected by the treatments. Table 5 shows the mean scores for the tested fabrics and the 7 pertinent attributes. For the silicone finishing, the slippery and greasy attributes change clearly with the concentration of the product. This result was expected as Ul treatment was known to soften the fabric and with the increase of concentration fabric becomes more greasy and slippery. It is also worth noting that the panel greatly perceived the modifications obtained by this treatment for the different concentrations.

For the resin treatment it is expected to have more responsive and less crumple-like fabrics. This is confirmed by the obtained results, since fabrics treated with a high concentration of resin finishing were significantly more nervous and less crumple-like than the non-treated fabric.

These results show that both treatments changed the hand-feel of the fabric in the expected direction and that the panel clearly perceived the modifications. Figure 14 shows the variation of sensory attributes according to the concentration of the finishing product.

The analysis of the results shows that the sensory evaluation ranges the treated fabrics as follows:

- for the resin finishing we have in terms of responsiveness 4<22<21<0, and for the crumple-like attribute 0<21<22<4;
- for the silicone treatment greasy and slippery attributes are ranged: 0<23<24<17.

Conclusion

The effects of finishing products' concentrations were found in accordance with the manufacturers' technical specifications and with the finishing industrialists' expectations. The evaluation of this effect was carried out by the sensory evaluation. The panel was able to detect the modifications and to evaluate them in the right sense.

Table 5. Mean values for the attributes according to the finishing treatments

	Non treated	K			UI		
Fabric code	0	21	22	4	23	24	17
Concentration	0	20	50	80	5	20	40
Falling	7.31	6.71	6.29	6.49	7.34	7.37	7.26
Rigid	3.09	3.90	4.01	4.38	3.18	2.78	2.99
Slippery	5.01	4.48	4.16	4.84	5.73	6.77	7.67
Soft	5.73	4.48	3.62	3.84	5.28	5.83	6.65
Greasy	2.14	1.81	1.55	1.70	2.98	4.77	5.52
Responsive	1.29	1.35	1.79	2.26	2.00	2.77	2.84
Crumple-like	7.60	6.98	6.06	4.47	7.32	7.67	7.40

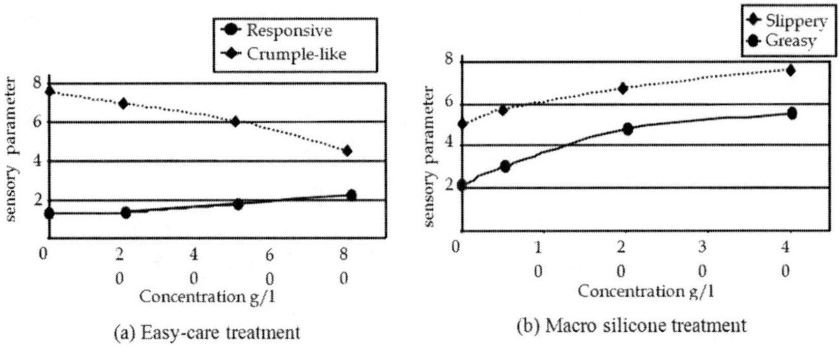

Figure 14. Variation of the effected attributes according to the concentration of the finishing product

6. CONCLUSION

Sensory analysis has become a powerful tool for helping textile industries in product design and marketing tasks. In fact, haptic perceptions, including both cutaneous and kinesthetic perceptions, guide consumers' choice for clothes as well as textile manufacturers for development of new products. Our studies on woven fabric have shown that modification of structure parameters or finishing treatments have a significant effect on sensory feeling. The trained panelists have detected those modifications. Sensory analysis methods provide quantification of tactile feeling. Moreover, sensory analysis approach allows understanding some complex sensation such as softness, comfort and well-being. It can therefore be concluded that sensory analysis has a solid future into the next century. In the meantime, development of dedicated devices for modeling of human perception and use of intelligent techniques which can be used in a complementary way for that purpose can be helpful and a promising approach.

REFERENCES

1. AFNOR 09-001 09001 (1983). Analyse sensorielle- Méthodologie- Directives générales

2. A. F. N. O. R. X. P. V., 09501 (1999). Sensory Analysis-General Guidance for Sensory Evaluation-Description, Differentiation and Hedonic Measurement

3. L. Bandini-Buti, L. Bonapace, A. Tarzia, 1997 Sensorial Quality Assessment: a method to incorporate perceived user sensations in product design. Applications in the field of automobiles. In IEA '97 Proceedings (Helsinki: Finnish Institute of Occupational Health), 186 9

4. J. Barthelemy, M. Danzart, Leod. P. Mac, F. Sauvageot, F. Sztrygler, 1990 Evaluation sensorielle. Manuel méthodologique, Ed. Technique et Documentation Lavoisier, Paris

5. S. Bensaid, J. Osselin-F, Schacher, L. , D. Adolphe, 2006 The effect of pattern construction on the tactile feeling evaluated though sensory analysis. Journal of the Textile Institute, 97 137 145

6. Said. S. Ben, L. Schacher, D. C. Adolphe, 2008 Touch and sight interaction in fabric sensory analysis" Tekstil, 57(8), 383 389

7. H. Binns, 1926 The discrimination of wool fabrics by the sense of touch. British Journal of Psychiatry, 16, 237 247

8. D. P. Bishop, 1996 Fabrics: Sensory and Mechanical Properties, Textile Progress, The Textile Institute 26

9. V. A. Cardello, C. Winterhalter, G. H. Schutz, 2003 Predicting the Handle and Comfort of Military Clothing Fabrics from Sensory and Instrumental Data: Development and Application of New Psychophysical Methods," Textile Research Journal, 73(3), 221-237

10. R. Chollakup, A. Sinoimeri, Philippe, F. Schacher, L. , Adolphe, 2004 (2004 a). Tactile sensory analysis applied to silk/cotton knitted fabrics. International Journal of Clothing Science and Technology, 16 132 140

11. R. Chollakup, A. Sinoimeri, F. Philippe, L. Schacher, Adolphe, 2004 (2004 b). Tactile feeling: sensory analysis applied to textile goods. Textile Research Journal, 74 1066 1072

12. C.. Cinel, G. W. Humphreys, R. Poli, 2002 Cross-Modal Illusory Conjunctions between Vision and Touch. Journal of Experimental Psychology: Human Perception and Performance, 28 5 1243 1266

13. F. Depledt, 1998 Société Scientifique d'Hygiène Alimentaire (SSHA) : Evaluation sensorielle-Manuel méthodologique, Lavoisier, Paris.

14. D. Dubois, H. Prade, 1997 Fuzzy criteria and fuzzy rules in subjective evaluation- a general discussion, Proceedings of EUFIT'97, Aachen, Germany, 975 979

15. H. M. Elder, S. Fisher, K. Armstrong, G. Hutchison, 1984 Fabric Softness, Handle and Compression, Journal of Textile Institute, 75 37 46

16. Jeguirim. S. El -Ghezal, Dhouib. A. Babay, M. Sahnoun, M. Cheikhrouhou, L. Schacher, D. Adolphe, 2009 The use of fuzzy logic and neural networks models for sensory properties prediction from process and structure parameters of knitted fabrics, Journal of Intelligent Manufacturing, Under press, DOI 10.1007/s10845-009-0362-y

17. Jeguirim. S. El -Ghezal, Dhouib. A. Babay, M. Sahnoun, M. Cheikhrouhou, L. Schacher, Adolphe, 2010 (2010 a). The tactile sensory evaluation of knitted fabrics: effect of some finishing treatments, Journal of the Sensory Studies, 25 2 April 2010, 201 215

18. Jeguirim. S. El -Ghezal, Dhouib. A. Babay, M. Sahnoun, M. Cheikhrouhou, L. Schacher, Adolphe, 2010 (2010 b). Sensory and instrumental techniques evaluating the effect of structure parameters on the tactile properties of knitted fabrics, Journal of Texture Studies, 41(5), 714- 735

19. Jeguirim. S. El -Ghezal, M. Sahnoun, Dhouib. A. Babay, M. Cheikhrouhou, L. Schacher, D. Adolphe, 2011 Predicting compression and surfaces properties of knits using fuzzy logic and neural networks techniques, International Journal of Clothing Science and Technology, Under press

20. S. Ertugrul, N. Ucar, 2000 Predicting bursting strength of cotton plain knitted fabrics using intelligent techniques. Textile Research Journal, 70 845 851

21. S. Fontaine, C. Marsiquet, N. Nicoletti, M. Renner, M. A. Bueno, 2005 Development of a sensor for surface state measurements using experimental and numerical modal analysis. Sensors and Actuators A, 120 507 517

22. J. Fortin, N. Durand, 2004 De la perception à la mesure sensorielle, La fondation des gouverneurs, Saint-Hyacinthe, Québec

23. A. Giboreau, S. Navarroa, P. Fayeb, J. Dumortier, 2001 Sensory evaluation of automotive fabrics: the contribution of categorization tasks and non verbal information to set-up a descriptive method of tactile properties. Food Quality and Preference 12 5-7 , July-September 2001, 311 322

24. S. Guest, C. Spence, (2003, (2003).What role does multisensory integration play in the visuotactile perception of texture? International Journal of Psychophysiology 50 63 80

25. S. Haykin, 2000 Neural Networks: A Comprehensive Foundation, Prentice Hall, New Jersey, 2nd edition, 1999. Presented at the ASME ICE Division Fall 2000 Technical Meeting September 25 27 , Peoria

26. H. S. E. C. A. New, to. Approach, Objective. the, of. Evaluation, Handle. Fabric, Mechanical. from, Part. I. Properties, Measure. Objective, Total. for, Evaluation. Handle, 2 ed., The Textile Machinery Society of Japan, Osaka, Japan, 1980, 7 28

27. C. L. Hui, T. W. Lau, S. F. Ng, 2004 Neural network prediction of human psychological perceptions of fabric hand, Textile Research Journal, 74(5), 375-383

28. M. Issa, L. Schacher, D. Adolphe, 2008 Development of a New Experimental Technique for Mechanical Characterization of Fabric, Experimental Techniques November/December, 24 29

29. ISO 1992ISO 5492 : (1992). Analyse sensorielle- Vocabulaire

30. I. S. O. 11035 19 ,. F. , 1995 Sensory analysis- Identification and selection of descriptors for establishing a sensory profile by a multidimensional approach

31. ISO 8586 International Standard ISO 8586-1993 (F), (1993), Assessors for Sensory Analysis, Part 1: Guide to the selection, training and monitoring of selected assessors

32. ISO 6658 International Standard ISO 6658-1985 (F), (1985), "Sensory analysis- Methodology- General guidance"

33. V. Jain, M. K. Tiwari, F. T. S. Chan, 2004 Evaluation of the supplier performance using an evolutionary fuzzybased approach. Journal of Manufacturing Technology Management, 15(8), 735-744

34. S. Kawabata, 1975 The standardisation and analysis of Hand Evaluation, Journal of the Textile Machinery Society of Japan', Osaka, Japan

35. S. Kawabata, 1980 The Standardisation and Analysis of Hand Evaluation (2nd Edition), Textile Machinery Society of Japan, Osaka, Japan

36. S. Kawabata, 19882 The Development of the Objective Measurement of Fabric Handle, Proceedings of the First-Japan Australia Symposium on Objective Specification of Fabric Quality, Mechanical Properties and Performance, Kyoto, Japan, 31 59

37. S. Kawabata, 1988 The Standardization and Analysis of Hand. Textile Research Journal, 438 444

38. M. Konyo, S. Tadokoro, M. Hira, T. Takamori, 2002 Quantitative Evaluation of Artificial Tactile Feel Display Integrated with Visual Information Proccedings of 2002 IEEE/RSJInternational Conference on Intelligent Robots and Systems EPFL, Lausanne October 2002 3060 3065

39. C. Kwak, A. Ventura, J. , K. Tofang-Sazi, 2000 A neural network approach for defect identification and classification on leather fabric, Journal of Intelligent Manufacturing, 11 485 499

40. S. J. Lederman, S. G. Abbott, 1981 Texture perception: Studies of intersensory organization using a discrepancy paradigm and visual versus tactual psychophysics. Journal of Experimental Psychology: Human Perception & Performance, 7(4), 902-915

41. S. J. Lederman, Thorne, G. , B. Jones, 1986 Perception of texture by vision and touch: Multidimensionality and intersensory integration. Journal of Experimental Psychology: Human Perception & Performance 12 169 180

42. N. Maâtoug, M. Sahnoun, F. Sakli, 2009 Banc d'essais pour la mesure des caractéristiques physiques d'état de surface des tricots. Tunisian Patent 19934, January, 12,.2009

43. C. Mackay, S. C. Anand, D. P. Bishop, 1999 Effects of laundering on the sensory and mechanical properties of 1x1 rib knitwear fabrics. Part II: Changes in sensory and mechanical properties, Textile Research Journal 69(4), 252 260.

44. T. Matsuo, N. Nasu, M. Saito, 1971 Study on the Hand, part 2: The Method for Measuring Hand, Journal of the Textile Machinery Society, 24(4), 58-68.

45. M. Meilgaard, G. Civille, B. Carr, 1991 Sensory evaluation techniques.- 2e éd., CRC Press Inc., Boca Raton, Floride, 354

46. A. Mucci, L. Garitta, G. Hough, S. Sampayo, 2005 Comparison of Discrimination Ability Between a Panel of Blind Assessors and a Panel of Sighted Assessors, Journal of Sensory Studies 20 28 34

47. M. Nagamachi, 1995 Kansei engineering: A new ergonomic consumer-oriented technology for product development. International Journal of Industrial Ergonomics, 15(1), 3-11

48. H. Nakano, 1994 Product Development of Clothes by Kansei Engineering. J. Soc. Fib. Sci. Tech Japan [Sen-i Gakkaishi] 50 [8], 473 478

49. NF EN 20139 1992. Textiles- Atmosphères normales de conditionnement et d'essai

50. NF-ISO 5492 (1992). Analyse sensorielle- Vocabulaire

51. NF-ISO 85861 (1993). Analyse sensorielle- Guide générale pour la sélection, l'entraînement et le contrôle des sujets- Partie1: sujets qualifiés

52. NF-ISO 11035 (1995). Analyse sensorielle- Recherche et sélection de descripteurs pour l'élaboration d'un profil sensoriel, par approche multidimensionnelle

53. H. Nicod, 1990 Evaluation Sensorielle, manuel méthodologique, SSHA, Technique and Documentation, Lavoisier, Paris, 2nd édition, 46 63Nogueira , C.; Cabeço-Silva, M. E.; Schacher, L. & Adolphe, D. (2009). Textile Materials: Tactile Describers. Journal of Food Technology 7(3): 66-70, 1684-8462

54. M. Okamoto, 1991 Some Attempts at Quantification of Sensibility" Toray Ind. Inc. J. Soc. Fib. Sci. Tech., Japan [Sen-i Gakkaishi] 47 [11], 617 623

55. M. J. Pac, M. A. Bueno, M. Renner, S. Elkasme, 2001 Warm- Cool Feeling Relative to Tribological properties of Fabrics, Textile Research Journal, 71 7 806 812

56. N. Pan, K. C. Yen, 1992 Physical Interpretations of Curves Obtained Through the Fabric Extraction Process for Handle Measurement. Textile Research Journal 62:279 290

57. N. Pan, H. Zeronian, H. S. Ryu, 1993 An Alternative Approach to the Objective Measurement of Fabrics, Textile Research Journal, 63 33 43

58. S. W. Park, Y. G. Hwang, B. C. Kang, 2000 Applying fuzzy logic and neural networks to total hand evaluation of knitted fabrics. Textile Research Journal, 70(8), 675 681

59. F. T. Peirce, 1930 The "Handle" of Cloth as a Measurable Quantity. J. Textile Inst. 21, T377

60. F. Philippe, L. Schacher, D. Adolphe, C. Dacremont, 2003 The sensory panel applied to textile goods: a new marketing tool, Journal of Fashion Marketing and Management 7, 235 248

61. D. Richard, D. Orsal, 2001 Neurophysiologie, Organisation et Fonctionnement du Système Nerveux, Dunod, Paris

62. F. Rombaldoni, R. Demichelis, G. Mazzuchetti, 2010 Prediction of human psychophysical perception of fabric crispness and coolness hand from rapidly measurable low-stress mechanical and thermal parameters. Journal of Sensory Studies, 25 2010 899 916

63. M. Rozenweig, A. Leiman, S. M. Breedlove, 1998 Psychobiologie, De Boeck Université, 978-2-74450-025-1 Paris, France

64. D. E. Rumelhart, G. E. Hinton, R. J. Williams, 1986 Learning Representations by Back-propagating errors, Nature 323, 533 536

65. P. Schlich, 1995 Preference Mapping: Relating Consumers Preferences to Sensory or Instrumental Measurements. Bioflavour, INRA Dijon, 135 150

66. Sensotact 2008 http: //www.sensotact.com

67. SSHA. 1998 Société Scientifique d'Hygiène Alimentaire : Evaluation Sensorielle- Manuel méthodologique. Technique & Documentation. 2-74300-124-0 Paris

68. H. Stone, J. L. Sidel, S. Oliver, A. Woolsey, R. C. Singleton, 1974 Sensory Evaluation by Descriptive Analysis. Food Technology, 28 11 24 34

69. H. Stone, J. L. Sidel, 2007 Sensory research and consumer-led food product development. In MacFie, H. (ed), Consumer-led Food Product Development. Boca Raton, FL: CRC Press; 9-07359-218-6 307 320

70. D. Tester, De Boos, 1990 1990. Get it right FAST time. Textile Horizons, 10(8), 13

71. S. Vassiliadis, M. Rangoussi, A. Cay, C. Provatidis, 2010 Artificial Neural Networks and Their Applications in the Engineering of Fabrics. Woven Fabric Engineering, Polona Dobnik Dubrovski, 111 134 , Sciyo, 978-9-53307-194-7 Croatia.

72. A. S. W. Wong, Y. Li, P. K. W. Yeung, P. W. H. Lee, 2003 Neural network predictions of human psychological perceptions of clothing sensory comfort. Textile Research Journal, 71, 331 337

73. W. K. Wong, C. K. Kwong, P. Y. Mok, W. H. Ip, 2006 Genetic optimization of JIT operation schedules for fabric-cutting process in apparel manufacture. Journal of Intelligent Manufacturing, 17, 341 354

74. N. D. Young, T. Sanders, M. A. Drake, J. Osborne, G. Civille, 2005 Descriptive analysis and US consumer acceptability of peanuts from different origins. Food quality and preference, 16 1), 37 43

75. L. A. Zadeh, 1965 Information and Control, 8, 338

76. X. Zeng, L. Koehl, M. Sahnoun, M. A. Bueno, M. Renner, 2004 Integration of human knowledge and measured data for optimization of fabric hand. International Journal of General Systems 33 (2-3), 243 258

77. X. Zeng, D. Ruan, L. Koehl, 2008 Intelligent sensory evaluation: Concepts, implementations and applications. Mathematics and Computers in Simulation, 77, 443 452

78. H. J. Zimmerman, 1996 Fuzzy Set Theory and Its Applications, 2nd Ed., Allied Publishers Limited, New Delhi

Chapter 10

Numerical prediction of in-plane permeability for multilayer woven fabrics with manufacture-induced deformation

Xuesen Zeng*, Andreas Endruweit, Louise P. Brown, Andrew C. Long

Faculty of Engineering, Division of Materials, Mechanics & Structures, University of Nottingham, University Park, Nottingham NG7 2RD, UK

ABSTRACT

A unit cell based Computational Fluid Dynamics model is presented for predicting permeability of multilayer fabric structures. In Liquid Composites Moulding processes, fabric lay-ups undergo significant manufacture-induced deformation, combining compression, shear, and inter-layer nesting. Starting from the configuration of un-deformed fabric, the deformation is simulated geometrically by accounting for self-imposed kinematic constraints of interweaving yarns. The geometrical modelling approach is implemented in the open-source software TexGen. The permeability tensor is retrieved from flow analysis in ANSYS/CFX, based on TexGen voxel models. Using only measured geometrical data for un-deformed fabrics, deformed plain weave fabric and twill weave fabric lay-ups were modelled and their permeability tensors predicted. Comparison with experimental data demonstrates the generally good accuracy of predictions derived from the proposed numerical method.

KEYWORDS

A. Fabrics/textiles; C. Computational modelling; E. Resin flow

1. INTRODUCTION

During preforming for Liquid Composite Moulding (LCM) processes, lay-ups of woven reinforcement fabric are subject to a combination of shear, compression and nesting. Previous studies have shown that the fabric permeability, which determines the subsequent impregnation with a liquid resin system, is highly sensitive to deformations of the fabric architecture due to shear [1], [2], [3], [4], [5] and [6], compression [7], [8] and [9], nesting [10], [11] and [12], and material handling history [13].

Experimentally determined fabric permeability data often show significant scatter and appear to depend on the experimental protocol, which was the focus of two recent international benchmark exercises [14] and [15]. The second benchmark exercise [15]confirmed that set-up and execution of experiments have an effect on measured permeabilities. Predictive permeability modelling has the potential to eliminate human errors, which occur in setting up and conducting experiments and may affect permeability data, provided that a suitable modelling approach is chosen.

In the literature, meso–macro approaches are proposed to describe the permeability as a function of the fibre volume fraction and the shear angle [1], [2], [3], [4] and [5]. Since details of the actual fabric structure are typically ignored, they need to be calibrated with experimental permeability values and are not truly predictive.

A number of numerical meso-scale modelling approaches for fabric permeability are documented in the literature [6], [16], [17], [18], [19], [20] and [21]. In some of these studies, the geometry of woven fabrics is idealized by assuming constant yarn cross-section and sinusoidal/zig-zag yarn paths. Consequently, the simplification in describing woven yarn geometries results in large differences between model predictions and experimental data, or the predictions remained un-validated. Other studies employ textile mechanics to predict realistic yarn geometries [6] and [19]. This requires mechanical properties of the textile, including shear and bending properties, to be characterised as input.

This study proposes a modelling approach, which is based on geometrical data and self-imposed kinematic constraints of interweaving yarns, to simulate woven fabric structures under realistic manufacturing conditions. The implementation is automated in the open source software TexGen [22] and [23] for generation of geometrical textile models. The commercial code ANSYS/CFX is employed for Computational Fluid Dynamics (CFD) analysis of impregnating flow through the generated fabric geometries as a pre-requisite for fabric permeability determination. The approach is evaluated for two woven fabric lay-ups at controlled compression and/or shear deformation.

2. GEOMETRICAL MODELLING OF FABRIC LAY-UPS

In a first step, realistic geometrical modelling is applied to single fabric layers. Moderate deformation of fabrics includes effects such as shear before reaching the locking angle and/or through-thickness compression of fabric lay-ups at low compaction pressure. In these cases, no excessive distortion of the fibre architecture occurs. The fabric geometry mainly undergoes kinematic changes, i.e. (1) re-orientation of yarn paths; (2) local rotation of yarn cross-sections around the yarn axis; (3) change of local yarn cross-sectional shapes. The deformation is simulated based on a purely geometrical modelling approach, not considering the actual yarn mechanics. For realistic modelling, basic geometrical considerations imply that the major axis of a rotated (elliptical) yarn cross-section is parallel to the tangent at the contact between two non-parallel (crossing) yarns. Artificial yarn intersections in the geometrical model are avoided through local adaptation of the yarn cross-sectional shape, which is applied automatically in TexGen and is based on conservation of the total yarn volume.

2.1. Fabric shear

Two identical idealised yarns, which were initially perpendicular and are now sheared by an angle γ, such that yarn II is now oriented at an angle $(90° - \gamma)$ relative to yarn I, are shown in Fig. 1(a) prior to any local rotation of yarn cross-sections around the yarn axis. The local rotational angle, θ, of the yarn cross-section of yarn II depends on the yarn cross-sectional width, w, yarn cross-sectional height, h, shear angle, γ, and nominal position, R, of the contact point. The normalised coordinate, R, with $0 < R < 1$ refers to a point on the segment of the axis of yarn II, where it crosses over yarn I (Fig. 1(a)). The sectional view A–A in Fig. 1(b), after applying the local rotation of yarn II around its axis, illustrates the tangential relationship of two elliptical cross-sections which determines the rotational angle, β. With the width $w' = w/\sin\gamma$, the parametric equations of the cross-section of yarn I in the A–A view are

$$x = \frac{w'}{2}\cos t = \frac{w}{2\sin\gamma}\cos t \text{ and } y = \frac{h}{2}\sin t, \tag{1}$$

where t is the eccentricity parameter ($0 \leqslant t \leqslant 2\pi$).

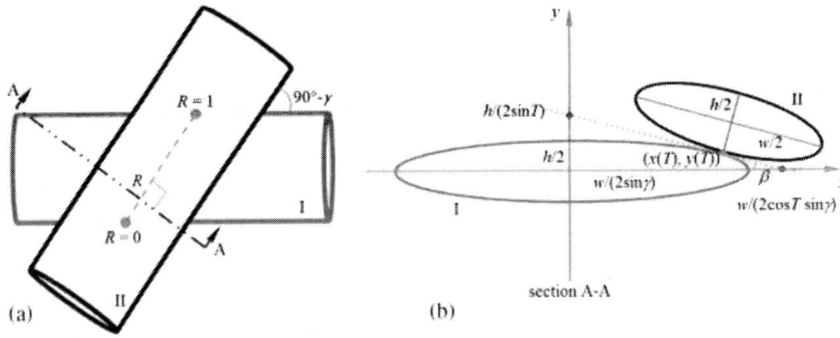

Figure 1. (a) Two identical elliptical cylinders, I and II, crossing over each other; (b) surface tangent between two ellipses at sectional view A–A of (a). (For interpretation of the references to colour in this figure legend, the reader is referred to the web version of this article.)

When the nominal contact point, at position R in Fig. 1(a), is projected along the y-axis in the A–A view, prior to cross-sectional rotation, onto the cross-section of yarn I, the projected point, $(x(T), y(T))$, is the actual contact point between cross-section I and II. Here, T is the eccentricity parameter of the contact point in the A–A section, which is related to R through

$$R = \frac{w'/2 - x(T)}{w'} = 0.5 - 0.5\cos T, \tag{2}$$

or

$$\cos T = 1 - 2R. \tag{3}$$

The tangent to the elliptical cross-sections at the contact point intersects the *x*- and *y*-axes at the points $(w/(2\cos T\sin \gamma), 0)$ and $(0, h/(2\sin T))$, respectively (Fig. 1(b)). These two intersection points are then used to calculate the local rotation angle, *β*, for the cross-section of yarn II. Local rotation angles are calculated for *R* varying from 0 to 1.

Based on the idealised derivation above, the tangential angle, *β*, would be approximately 90° at contact points near $R = 0$ and $R = 1$. This would introduce an unrealistic local rotation of the yarn cross-section. Hence, to avoid the extreme 90° rotation, the rotational angle is calculated by offsetting the contact point position, *R*. Implementation of the offset scheme in TexGen allows realistic local yarn rotations at varying contact points ($0 < R < 1$) to be captured and values for *β* near 90° to be avoided near $R = 0$ and $R = 1$ (Fig. 2). The following offset scheme is applied in TexGen

$$R' = \begin{cases} R + \frac{1}{10}\sin\gamma & \text{if } 0.0 \leqslant R < 0.5 \\ R - \frac{1}{10}\sin\gamma & \text{if } 0.5 < R \leqslant 1.0 \end{cases} \tag{4}$$

The offset, $1/10 \sin \gamma$, reflects the finite compliance of yarn II in conforming to the surface of yarn I. Based on the assessment of TexGen model for a range of offset values, the factor, 1/10, was found to give plausible results in terms of yarn curvature for a wide range of yarn cross-sectional shapes. The discontinuity at $R = 0.5$ in Eq. (4) is avoided through interpolation of the yarn path based on 20 discrete data points sampled by TexGen. With severe inter-yarn intersections being avoided in a unit cell model, the procedure of applying local yarn rotation (Eqs. (1), (2), (3) and (4)) effectively represents the main yarn geometry changes during shear of a woven fabric.

2.2. Yarn cross-section variation under shear and compaction

Rotation of the yarn cross-section (Section 2.1) prevents severe intersection of yarns at the cross-over point, illustrated in Fig. 3(a) and (b). For any remaining interference between yarns, further changes are made to the yarn cross-sectional shape: When intersections between a yarn cross-section and an adjacent yarn are detected, the positions of points on the cross-section are adjusted, replacing the originally elliptical cross-section with a polygon. The resulting yarn geometry is shown in Fig. 3(c).

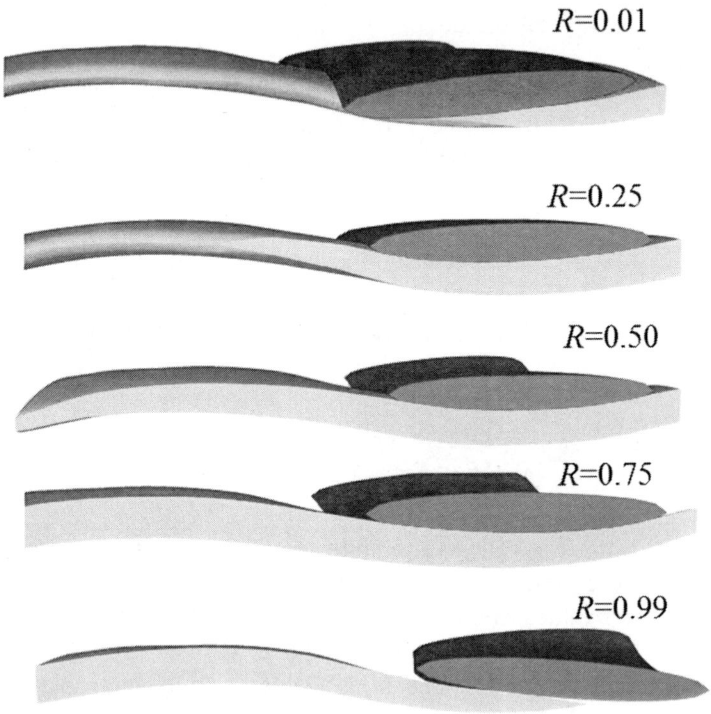

Figure 2. Predicted yarn rotations along the axis of the top yarn (different values of *R*) in a TexGen model. (For interpretation of the references to colour in this figure legend, the reader is referred to the web version of this article.)

It has been observed experimentally that in compaction of fabric lay-ups, yarns tend to flatten, and there is an offset between the axis of one yarn and the point at which the yarn passing over it drops away (Fig. 4). In order to simulate the geometrical features presented in Fig. 4, an algorithm has been developed which allows the yarn cross-sectional shape to be varied along the yarn axis. It is assumed that each yarn is bound between two parallel planes at a distance corresponding to the specified fabric thickness (Fig. 5). For each cross-section, the nearest distances between points on the yarn perimeter and either the upper or lower bounding plane, $\Delta\tau$, are identified. If $\Delta\tau < 0$ for any point on the yarn perimeter, there are intersections of the yarn with the nearest bounding plane. The yarn cross-sectional shape is corrected by moving the points until the respective values of $\Delta\tau$ are zero, thus constraining the yarn to the specified fabric thickness. For points on the yarn perimeter with initial values $\Delta\tau > 0$, the position is adjusted using an iterative method to restore the original cross-sectional area whilst being bound within the planes. Fig. 3(d) shows the final stage of the refinement, where the height of the upper yarn is increased until it reaches the limit determined by the overall fabric thickness. The side view of the lower yarn shows the yarn drop offset produced by the height adjustment at different sections along the yarn.

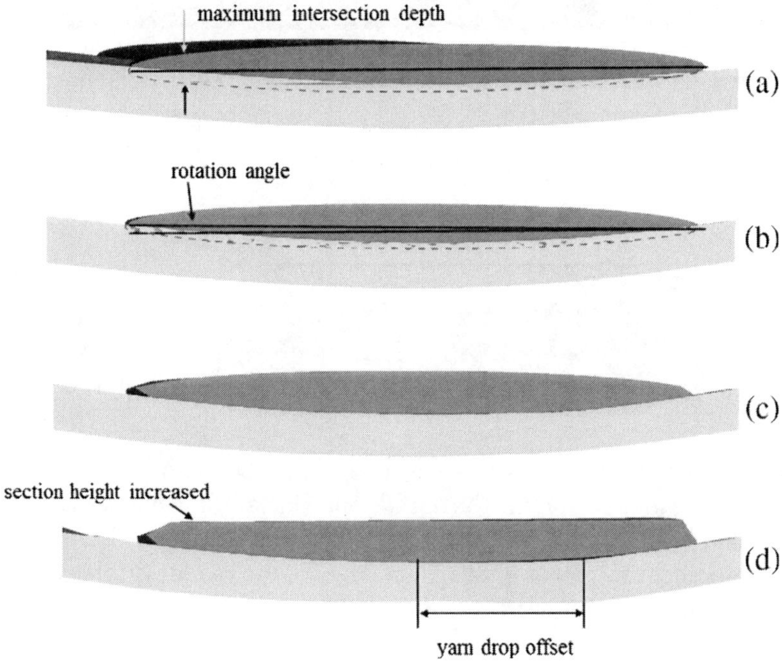

Figure 3. Geometrical refinement stages: (a) unrefined; (b) rotation only; (c) rotation and intersection correction; (d) rotation, intersection correction and adjusted section height. (For interpretation of the references to colour in this figure legend, the reader is referred to the web version of this article.)

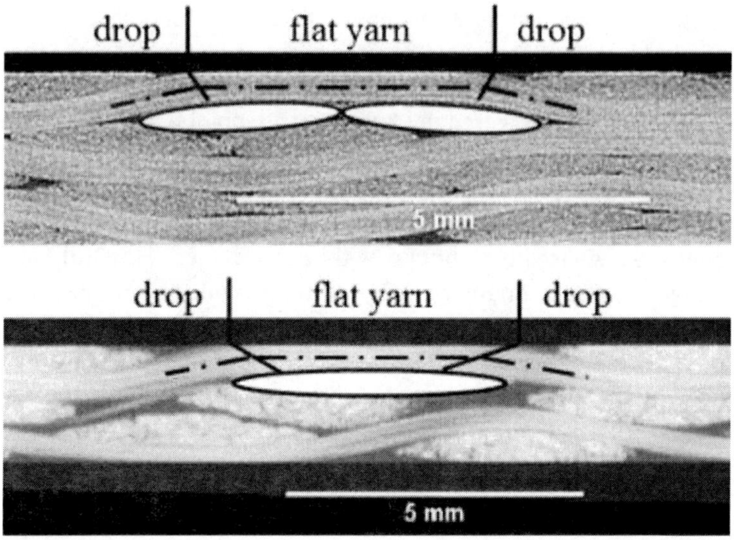

Figure 4. Yarn drop offset under compression in twill weave fabric (top) and plain weave fabric (bottom).

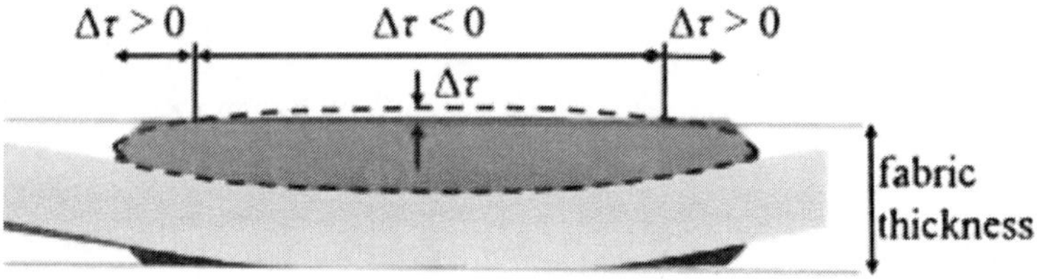

Figure 5. Intersection of yarn cross-section with bounding planes for fabric thickness, with penalty, $\Delta\tau$. The dashed line indicates the initial shape whilst the solid colour is the final shape after applying $\Delta\tau$. (For interpretation of the references to colour in this figure legend, the reader is referred to the web version of this article.)

Based on the refinement approach above, cross-sections along the path of a single yarn vary systematically depending on the interweaving geometrical constraints. Fig. 6compares simulated yarn cross-sectional variations with those in the real material. The figure suggests that, despite the simplifications in the modelling process, the main features of real yarn cross-sections, such as asymmetry, varying aspect ratio, and distinct tips, are reproduced in the model.

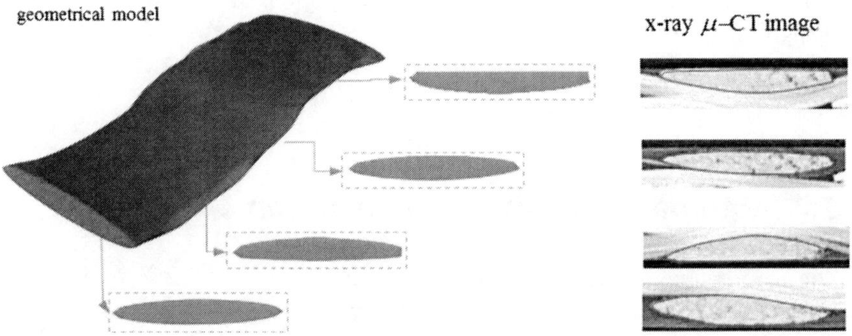

Figure 6. Yarn cross-sectional details in an automatically generated geometrical model, representing typical features found in X-ray μ-CT scans of the corresponding plain weave fabric laminate composites. (For interpretation of the references to colour in this figure legend, the reader is referred to the web version of this article.)

2.3. Fabric lay-up

Previous studies have highlighted the influence of nesting in multilayer reinforcements on permeability scatter [11] and [12]. Nesting involves complex interactions between fabric layers leading to layer thickness changes and localised compaction. Since the interdependence between in-plane layer shift and through-thickness compaction cannot be modelled accurately, the effect of nesting on the formation of gap spaces in the lay-up is approximated in two stages. In the first stage, each individual fabric layer undergoes compaction between two (virtual) flat mould surfaces, the spacing of which corresponds to the average layer thickness. In the second stage, multiple (identical) layers of the compacted flat fabric are stacked on top of each other, and in-plane shifting in the warp and weft direction is applied. Although this approach does not capture the precise interaction between layer shift and local compaction, it reflects the approximate geometry and connectivity of flow channels in between fabric layers at given global fibre volume fraction.

Latin hypercube sampling [24] was employed to generate in-plane shift vectors. Compared to generic Monte-Carlo sampling, Latin hypercube sampling produces more evenly distributed samples from the entire ranges of the variables, thus allowing to fully explore multi-dimensional inputs based on limited numbers of simulations [25] and [26]. The Latin hypercube design function *X* = lhsdesign(*N*, *P*) from MatLab – Release 2013 was used, which generates a sample *X* containing *N* values on each of *P* variables.

2.4. Automation of unit cell generation

The modelling approach described in Sections 2.1, 2.2 and 2.3 is implemented and automated as a 2D fabric wizard in TexGen. Fig. 7 illustrates unit cell models of a single ply of a plain weave fabric at various shear angles. Fig. 8 shows two lay-up models, 3 layers of a sheared plain weave fabric and 9 layers of a twill weave fabric. The unit cell domain is defined as shown in Fig. 8 to ensure the translational periodicity for each fabric layer at any shear angle and in-plane shift. Periodicity is utilized later in the CFD analysis described in Section 3. The geometrical unit cell models are discretised into voxels and converted to input files for flow analysis in ANSYS/CFX.

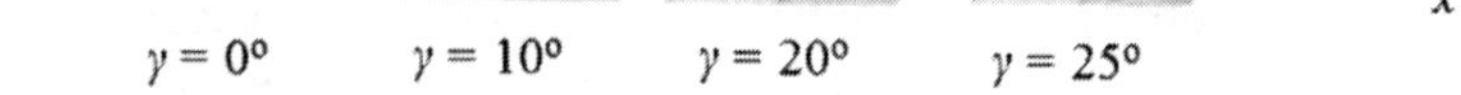

Figure 7. Automatically generated unit cell models with shear angles of 0°, 10°, 20° and 25° relative to the *y*-axis.

3. CFD ANALYSIS FOR PERMEABILITY PREDICTION

The meshed CFD models comprise yarns and the surrounding empty spaces in the mould cavity for resin injection. The permeabilities of yarns, which can be estimated based on the models proposed by Gebart [27], assuming ideal square or hexagonal filament arrangement, are typically two orders smaller than the overall permeabilities of the fabric lay-ups. Thus, the yarns can be assumed to be impermeable which helps to reduce the computational cost for CFD simulations. This assumption was validated numerically by Nedanov and Advani [16]. Steady-state flow through the pore spaces was simulated. Translational periodic boundary conditions were applied on opposite faces of the textile unit cell domain in weft and warp directions to represent a continuous reinforcement. A flow-driving pressure drop was applied in either warp or weft direction. Non-slip wall boundary conditions were specified at the top and bottom faces of the domain to simulate flow along the mould surfaces during in-plane fabric impregnation. Since inter-yarn gap spaces are typically two orders larger than pore spaces in the yarns, lubrication at the yarn-gap interface can be assumed to be negligible, based on the study by Grouve and Akkerman [28]. Hence non-slip boundary conditions were also applied on the yarn surfaces. To validate the assumptions for flow modelling, CFD simulations were run for single layers of the two fabrics analysed in this study, where the porosity of yarns was considered, and finite slip was applied on the yarn surfaces. Permeabilities predicted based on the method described in the following differed by less than 1% from corresponding values solid yarns and non-slip boundary conditions.

Darcy's law for a 2D flow problem can be expressed in global x–y coordinates as

$$\begin{pmatrix} u \\ v \end{pmatrix} = \frac{1}{\eta}\begin{pmatrix} K_{xx} & K_{xy} \\ K_{yx} & K_{yy} \end{pmatrix}\begin{pmatrix} \partial p/\partial x \\ \partial p/\partial y \end{pmatrix}, \qquad (5)$$

where u and v are the flow velocity components, K_{xx}, K_{yy}, and $K_{xy} = K_{yx}$ are the components of the permeability tensor to be determined by unit cell based CFD analysis, p is the fluid pressure, and η is the fluid viscosity. For a sheared unit cell domain, translational periodic boundary conditions are applied on the pairs of face A/face B and face C/face D (Fig. 7). From each CFD simulation, a flow velocity vector is calculated on faces A, B, C and D as averaged nodal velocity weighted by the area of the respective face. The permeability tensor in Eq. (5) is determined based on two flow cases.

In case I, a pressure drop, Δp, is imposed between face C and face D (the distance between which is H), such that $\partial p/\partial x = 0$ and $\partial p/\partial y \approx \Delta p/H$. With the resulting velocity components, u_1 and v_1, Eq. (5) can be solved:

$$K_{xy} = K_{yx} = \frac{u_1 \eta H}{\Delta p};\ K_{yy} = \frac{v_1\ \eta H}{\Delta p} \tag{6}$$

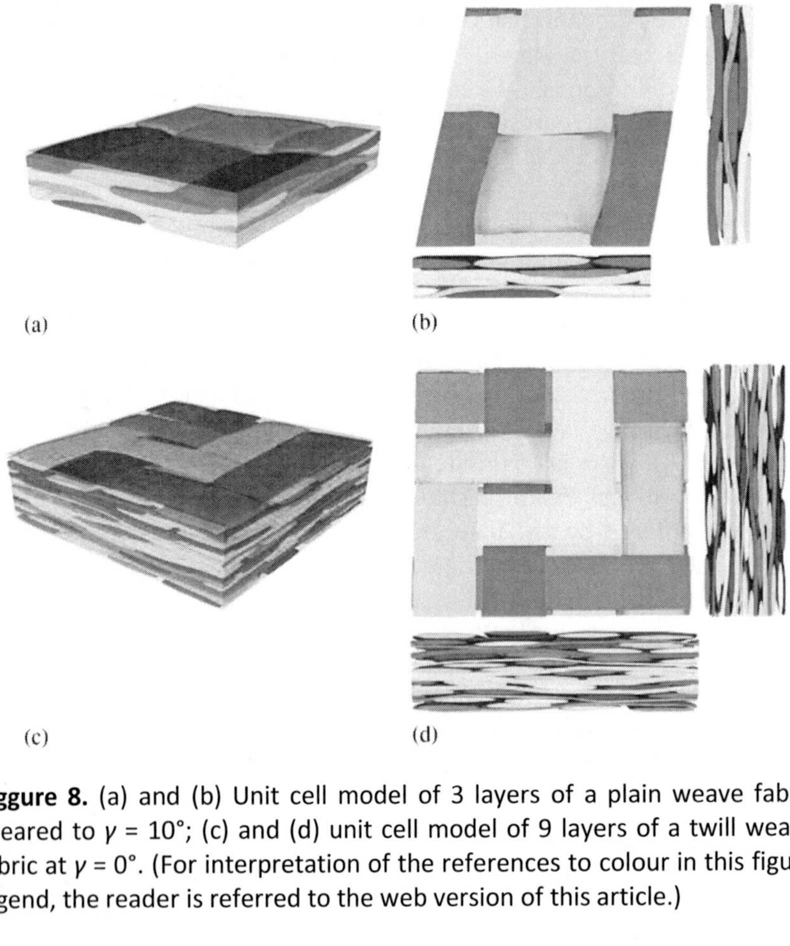

Figgure 8. (a) and (b) Unit cell model of 3 layers of a plain weave fabric sheared to γ = 10°; (c) and (d) unit cell model of 9 layers of a twill weave fabric at γ = 0°. (For interpretation of the references to colour in this figure legend, the reader is referred to the web version of this article.)

In case II, a pressure drop, Δp, is imposed between face A and face B (the distance between which along the x direction is L), such that $\partial p/\partial x \approx \Delta p/L$ and $\partial p/\partial y \approx \Delta p/(L \tan \gamma)$, where γ is the shear angle. With u_2, v_2 and Eq. (5):

$$K_{xx} = \frac{\eta L}{\Delta p}(u_2 - \frac{H \tan\ \gamma}{L} u_1) \tag{7}$$

The permeability tensor is then transformed to the diagonal form (K_1 and K_2) [29], to allow direct comparison with experimental data.

4. MATERIALS AND PERMEABILITY MEASUREMENT

4.1. Materials

A 2 × 2 twill weave carbon fibre fabric (with a nominal areal density S_0 = 285 g/m^2 and filament count c_f = 6 K) and a plain weave glass fibre fabric (with S_0 = 895 g/m^2) were selected for validating the numerical approach for permeability prediction. The two fabrics differ in weave style and areal density, which leads to distinctive features in meso-scale yarn geometry, as shown in Fig. 9. These image data were acquired from two sources: (a) 3D X-ray micro-Computed Tomography (μ-CT) using a GE Phoenix Nanotom 180NF system and (b) 2D flatbed document scanner. The μ-CT image data offer three-dimensional spatial information on the fabric geometry. However, the size of the scan volume needs to be limited in order to allow high resolution image data to be acquired. For example, the composite specimen with 9 layers of twill weave fabric was scanned in dimensions of 12 mm × 12 mm × 3.5 mm at a resolution of 6.75 μm/voxel, whilst the specimen with 3 layers of the plain weave fabric was scanned at 30 mm × 30 mm × 2 mm with a resolution of 20 μm/voxel.

To assess the statistical variability of woven fabric geometry over a large number of weave repeats [30] and [31], complementary to μ-CT data, 2D images were acquired using the flatbed scanner providing a large surface view of the fabric (210 mm × 297 mm) at a resolution of 21.17 μm/pixel. This enables statistically robust measurement of yarn width and yarn spacing. To quantify each of these geometrical parameters, 120 measurements were taken. Average values, standard deviations and coefficients of variation are listed in Table 1.

4.2. In-plane permeability measurement

For the 2 × 2 twill weave fabric, in-plane permeability data were acquired in the framework of the recent permeability benchmark II [15], where use of an experimental set-up for unsaturated linear flow at constant injection pressure was prescribed for permeability characterisation of the unsheared fabric. For the experimental set-up used here, the relative experimental error for each individual experiment, determined by uncertainties in measurement of flow front propagation, fibre volume fraction, fluid viscosity, and applied injection pressure, was quantified as ±7%. At a cavity height of 3.5 mm, 9 fabric layers resulted in a fibre volume fraction V_f = 0.40.

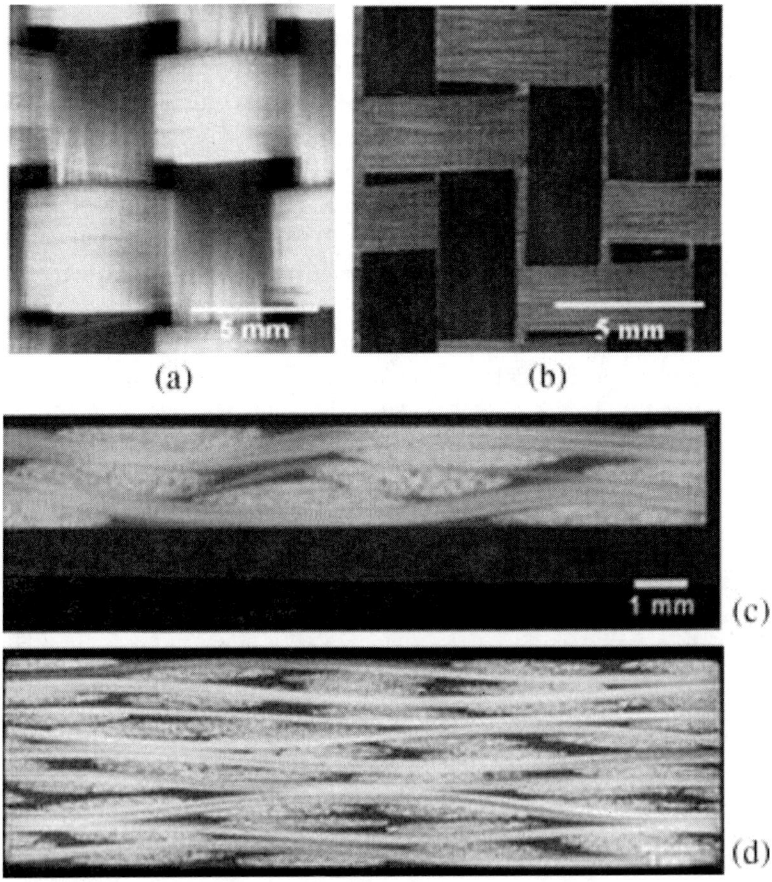

Figure 9. (a) Glass fibre plain weave fabric; (b) carbon fibre twill fabric; (c) μ-CT image of 3-layer laminate made from the fabric in (a); (d) μ-CT image of 9-layer laminate made from the fabric in (b).

Table 1. Geometry parameters determined from 3D μ-CT scans of laminates and from 2D scans of single fabric plies: inter-yarn gap width, *g*, yarn width, *w*, and yarn height, *h*, in both fabric directions; average values, standard deviations and coefficients of variation (standard deviation/average) are given where appropriate.

		$K_1/10^{-10}\,m^2$	$K_2/10^{-10}\,m^2$	θ (°)	K_1/K_2
9-layer twill weave ($\gamma = 0°$, $V_f = 0.40$)	Experiment	3.65 ± 0.44 (±12%)	2.24 ± 0.25 (±11%)	1 ± 14	1.65 ± 0.29 (±18%)
	Prediction	3.96 ± 0.51 (±13%)	2.50 ± 0.18 (±7%)	1 ± 5	1.59 ± 0.22 (±14%)
3-layer plain weave ($\gamma = 0°$, $V_f = 0.51$)	Experiment	2.29 ± 0.76 (±33%)	1.41 ± 0.76 (±54%)	19 ± 37	1.74 ± 0.39 (±22%)
	Prediction	5.67 ± 1.39 (±25%)	0.80 ± 0.17 (±21%)	−1 ± 1	7.68 ± 3.43 (±45%)
3-layer plain weave ($\gamma = 10°$, $V_f = 0.52$)	Experiment	1.63 ± 0.43 (±26%)	1.04 ± 0.27 (±26%)	−14 ± 28	1.57 ± 0.03 (±2%)
	Prediction	2.10 ± 0.85 (±40%)	0.94 ± 0.60 (±64%)	−14 ± 15	3.72 ± 1.11 (±30%)
3-layer plain weave ($\gamma = 20°$, $V_f = 0.55$)	Experiment	1.32 ± 0.92 (±70%)	0.33 ± 0.17 (±51%)	−2 ± 6	3.80 ± 0.90 (±24%)
	Prediction	1.20 ± 0.24 (±20%)	0.51 ± 0.18 (±35%)	−22 ± 10	4.61 ± 6.24 (±17%)
3-layer plain weave ($\gamma = 25°$, $V_f = 0.56$)	Experiment	0.80 ± 0.22 (±27%)	0.28 ± 0.07 (±27%)	−11 ± 10	2.88 ± 0.43 (±15%)
	Prediction	0.92 ± 0.35 (±38%)	0.26 ± 0.09 (±35%)	−17 ± 16	2.66 ± 1.10 (±41%)

For the plain weave fabric, the in-plane permeability was measured in unsaturated radial flow experiments at constant injection pressure [32]. Due to the layout of radial injection experiments, sheared fabric specimens can be accommodated. For the set-up used here, the relative experimental error for each individual experiment was ±14%. At a cavity height of 2 mm, three fabric layers corresponded to V_f = 0.51 for the case of non-sheared fabric (γ = 0°). Additional permeability measurements were made on 3 layers of the sheared plain weave fabric at a constant cavity height of 2 mm. The resulting fibre volume fractions were V_f = 0.52 at a shear angle γ = 10°, V_f = 0.55 at γ = 20°, andV_f = 0.56 at γ = 25°. The uncertainty in shearing the fabric specimens to given angles before they were placed in the injection tool cavity is estimated as ±5°.

5. RESULTS AND DISCUSSIONS

5.1. Evaluating permeability predictions

In order to compare numerical predictions with experimental measurements (Section4.2), unit cell models of the tested fabric lay-ups were generated in TexGen using the geometrical data for un-deformed fabrics (2D scans) listed in Table 1 as input. Six stochastic models with varying in-plane layer shift (Section 2.3) were created for each lay-up. To study the mesh sensitivity of CFD analysis based on these models, the following numbers of voxel elements for the unit cell were tested for 9 layers of the twill weave fabric (Fig. 8(c) and (d)): 50 × 50 × 90 (warp × weft × thickness), 100 × 100 × 180, 150 × 150 × 270 and 300 × 300 × 270. Convergence of the permeability predictions was found for a mesh with 150 × 150 × 270 voxels. Similarly, for 3 layers of the plain weave fabric (Fig. 8(a) and (b)), results converged for a mesh of 150 × 150 × 90 voxels.

The numerical results for in-plane permeabilities are listed in Table 2 together with the corresponding experimental data. In general, the difference between average values of predicted and experimentally determined permeability is in the same order or smaller than the standard deviation of either value. For the twill weave fabric lay-up, the predictions of the permeability tensor (K_1 and K_2) deviate on average by 10% from the experimental data. For the plain weave fabric lay-up, the average deviation of the predicted values for K_1 and K_2 from the experimental values is 20% when sheared (γ = 10°, 20° and 25°). Here, the prediction for the un-sheared plain weave fabric lay-up (γ = 0°) is an outlier deviating from the experimental data by 150% for K_1, and by 30% forK_2.

Table 2. Experimental data and CFD predictions of in-plane permeability for different shear angles, γ: principal permeability values, K_1 and K_2, and angle between K_1 and weft direction, ϑ; average values, standard deviations and coefficients of variation (standard deviation/average) are given where appropriate.

		$K_1/10^{-10}\,m^2$	$K_2/10^{-10}\,m^2$	ϑ (°)	K_1/K_2
9-layer twill weave ($\gamma = 0°$, $V_f = 0.40$)	Experiment	3.65 ± 0.44 (±12%)	2.24 ± 0.25 (±11%)	1 ± 14	1.65 ± 0.29 (±18%)
	Prediction	3.96 ± 0.51 (±13%)	2.50 ± 0.18 (±7%)	1 ± 5	1.59 ± 0.22 (±14%)
3-layer plain weave ($\gamma = 0°$, $V_f = 0.51$)	Experiment	2.29 ± 0.76 (±33%)	1.41 ± 0.76 (±54%)	19 ± 37	1.74 ± 0.39 (±22%)
	Prediction	5.67 ± 1.39 (±25%)	0.80 ± 0.17 (±21%)	−1 ± 1	7.68 ± 3.43 (±45%)
3-layer plain weave ($\gamma = 10°$, $V_f = 0.52$)	Experiment	1.63 ± 0.43 (±26%)	1.04 ± 0.27 (±26%)	−14 ± 28	1.57 ± 0.03 (±2%)
	Prediction	2.10 ± 0.85 (±40%)	0.94 ± 0.60 (±64%)	−14 ± 15	3.72 ± 1.11 (±30%)
3-layer plain weave ($\gamma = 20°$, $V_f = 0.55$)	Experiment	1.32 ± 0.92 (±70%)	0.33 ± 0.17 (±51%)	−2 ± 6	3.80 ± 0.90 (±24%)
	Prediction	1.20 ± 0.24 (±20%)	0.51 ± 0.18 (±35%)	−22 ± 10	4.61 ± 6.24 (±17%)
3-layer plain weave ($\gamma = 25°$, $V_f = 0.56$)	Experiment	0.80 ± 0.22 (±27%)	0.28 ± 0.07 (±27%)	−11 ± 10	2.88 ± 0.43 (±15%)
	Prediction	0.92 ± 0.35 (±38%)	0.26 ± 0.09 (±35%)	−17 ± 16	2.66 ± 1.10 (±41%)

For this case ($\gamma = 0°$), the interference between yarns is severe, as shown in Fig. 10(a) prior to any refinement. After automated geometrical refinement to correct the interference whilst keeping the fibre volume fraction constant, the resulting model (Fig. 10(b)) has a high degree of resemblance to the real material (Fig. 10(c)). However, the sharp taper of tips in the yarn cross-sections, highlighted with arrows in Fig. 10(b), is recognised as being artificial in comparison with the μ-CT scan. It results from the resolution of cross-sectional modelling in TexGen, which is limited by the number of points defining the yarn perimeter. This geometrical detail introduces additional flow channels in the numerical model for $\gamma = 0°$, leading to enhanced flow in the simulation and over-prediction of the permeability. Nonetheless, Fig. 10(d) shows that the significant difference between predicted and experimental values for K_1 is reduced for models with stochastic layer shift and refined yarn geometry (Fig. 10(b)) compared to stochastic models without any geometrical refinement (Fig. 10(a)).

For the unit cell models of sheared plain weave fabrics, the artificial taper of cross-sectional tips highlighted in Fig. 10(b) is avoided as illustrated in Fig. 2 (R = 0.25, 0.50 and 0.75), which is helped mainly by modelling the yarn cross-sectional rotation due to shear (Section 2.1). As a result, the permeability predictions are more accurate for the sheared plain weave fabrics than for the un-sheared fabric.

In general, the numerical predictions achieved higher accuracy for the twill weave fabric lay-up than for the plain weave fabric lay-ups, which can be explained by the higher geometrical complexity in modelling tightly packed woven architectures related to the presence of more geometrical constraints. Since the twill weave fabric ($V_f = 0.40$) has lower fibre volume fraction than the plain weave fabric ($V_f = 0.51$), less surface interference occurs in the geometrical unit cell model prior to any refinement. The lower the degree of surface interference, the more realistic is the geometrical model resulting from the refinement procedure (Section 2), and the more accurate is the permeability prediction.

5.2. Effect of shear on permeability

Numerical predictions of the permeability tensor for the plain weave fabric were obtained as a function of the shear angle, γ. For comparison, a simple analytical model, which was originally proposed by Long et al. [33], Smith et al. [2] and recently applied by Aranda et al. [1], is adapted to relate the in-plane permeability, K_1 and K_2, to the shear angle, γ. The model is based on a Kozeny–Carman type dependence on the fibre volume fraction and is expressed as

$$\begin{pmatrix} K_1 \\ K_2 \end{pmatrix} = \begin{pmatrix} A_1 \\ A_2 \end{pmatrix} \frac{(1-V_f)^3}{V_f^2} \cos^2\left(\frac{90^\circ + \gamma}{2}\right), \quad (8)$$

where A_1 and A_2 are derived from the experimentally determined permeability of the un-sheared fabric laminate (at $\gamma = 0°$) and assumed to be constant. The fibre volume fraction, V_f, is a function of γ.

The angle ϑ is defined as the angle between the direction of the principal permeability, K_1, and the weft yarns. An empirical equation between ϑ and the shear angle, γ, proposed by Endruweit and Ermanni [5] is used:

$$\theta = \frac{90^\circ - \gamma}{2}\left(1 - \sin^x(90^\circ - \gamma)\right), \text{ where } x = 5\left(\left(\frac{K_1}{K_2}\right)_{\gamma=0^\circ} - 1\right)^{-1} \quad (9)$$

Fig. 11 compares the analytical/empirical model in Eqs. (8) and (9) with CFD predictions and experimental data. For K_1 and K_2 (Fig. 11(a) and (b)), the decrease in permeability with increasing shear angle, indicated both by the CFD results and Eq. (8), is a result of superposition of the change in fibre volume fraction (described by the Kozeny–Carman equation) and the re-orientation of flow channels. This view is consistent with previous studies [1], [2] and [5]. The numerical results predict experimental data generally with reasonable accuracy and follow the same trend as formulated in the analytical model in Eq. (8). Unlike the analytical model, the CFD simulations do not require calibration by experimental data of the un-sheared fabric ($\gamma = 0°$). Hence, the numerical approach has the advantage of being a truly predictive tool. Fig. 11(c) indicates that CFD predictions of the angle ϑ are in reasonable agreement with experimental data and the empirical model (Eq. (9)), which was deduced from experimental permeability data for different materials [5]. The data listed in Table 2 suggest that the scatter of the angle ϑ is related to the ratio of the major and minor axes of the flow front ellipse forming in radial flow, R_1/R_2, which is equal to the square root of the ratio K_1/K_2. The smaller the average K_1/K_2, the more sensitive is the orientation of the flow front ellipse to small changes in the lay-up structure, e.g. caused by the random in-plane shift of fabric layers, which is discussed in the following section.

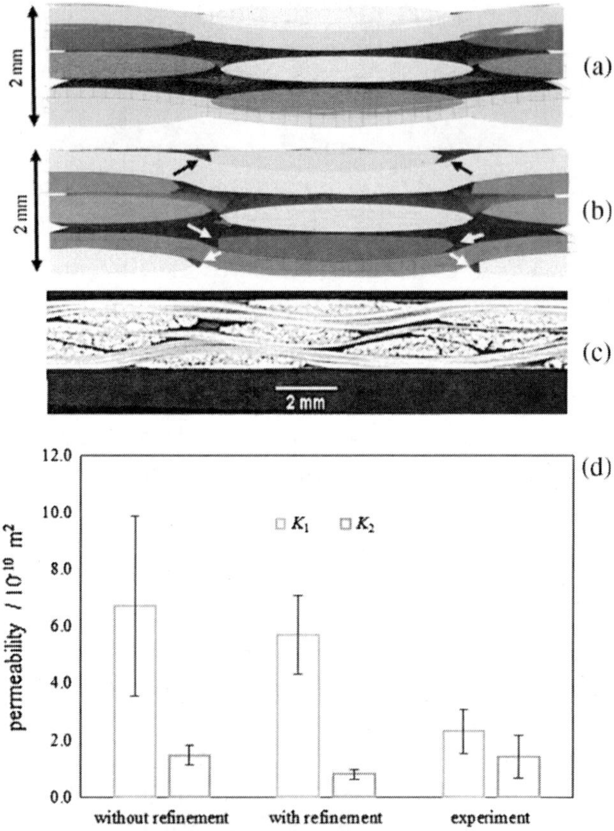

Figure 10. Three layers of plain weave fabric: (a) TexGen model prior to any refinement to correct yarn intersections; (b) after the refinement; (c) μ-CT image of laminate cross-section; (d) CFD predictions based on six stochastic models and experimental results for K_1 and K_2. (For interpretation of the references to colour in this figure legend, the reader is referred to the web version of this article.)

5.3. Prediction of permeability scatter

Characterising and predicting the variability of reinforcement permeabilities is a pre-requisite for robust design of LCM processes. For the fabric lay-ups in this study, standard deviations of experimental data and numerical predictions are listed in Table 2. Comparing data for unsheared fabrics, the permeability scatter for 9 layers of the twill weave fabric (experiment: up to ±12%; prediction: up to ±13%) is lower than for 3 layers of the plain weave fabric (experiment: up to ±54%; prediction: up to ±25%). With increasing number of layers, the effect of lay-up randomness on the permeability decreases. As reported by Lomov et al. [11], the distribution of the average thickness per layer becomes narrower, and the average nesting configuration tends to converge when the number of layers increases. In addition, the influence of variations in the structure of individual layers (in actual specimens) on the permeability of the lay-up decreases with increasing number of layers. Numerical modelling, as illustrated in Fig. 12, enables a direct relation between the lay-up randomness, the flow channel distribution and the scatter of the permeability tensor to be derived. The flow channel distributions in Fig. 12were visualised in ANSYS CFD-Post. For 3 layers of the plain weave fabric (γ = 25°), the random in-plane shift of layers in warp and weft direction (Fig. 12(a), top) merges or separates the flow channels at each interface between layers. It is clear in Fig. 12(a) that model 6 has the most interconnected flow channels, whilst model 3 shows most isolated flow channels between fabric layers. The corresponding values of K_1 are 1.47×10^{-10} m^{2}for model 6 and 0.77×10^{-10} m^2 for model 3 as show in Fig. 12(b). Similarly, K_2 for model 6 is 2.5 times that for model 3. The angle ϑ for model 3 is −37°, indicating that the orientation of K_1 rotates towards the warp direction. For model 6, ϑ = −9° is closer to the weft direction.

6. CONCLUSION

This study presents a numerical approach for predicting permeabilities of deformed fabrics, which is based on geometrical data for un-deformed fabrics as input. Permeability values derived from simulations of impregnating flow through fabric structures are not affected by operator-induced experimental issues and have the potential to be more reliable than experimental data, provided that numerical modelling is accurate. Starting from the un-deformed fabric geometry, two levels of variation in material structure are modelled in TexGen to simulate realistic deformations of the lay-up material. The first level is related to yarn shape variation due to geometrical constraints imposed by shear and compression. The second level is related to stochastic variability in ply stacking. Comparison of permeability values predicted based on flow simulations using these detailed geometrical models with corresponding experimental data indicated an average difference of 10% for a twill weave fabric lay-up at $V_f = 0.40$ and an average difference of 20% for a plain weave fabric lay-up at different shear angles with $V_f > 0.52$. Considering that, unlike for existing analytical models, no calibration factor is required, the predictions show good accuracy, which also reflect that the refinement in geometrical modelling is more accurate for fabrics with low V_f. In addition, this study demonstrates that permeability scatter is closely related to the meso-scale geometry of fabric lay-ups.

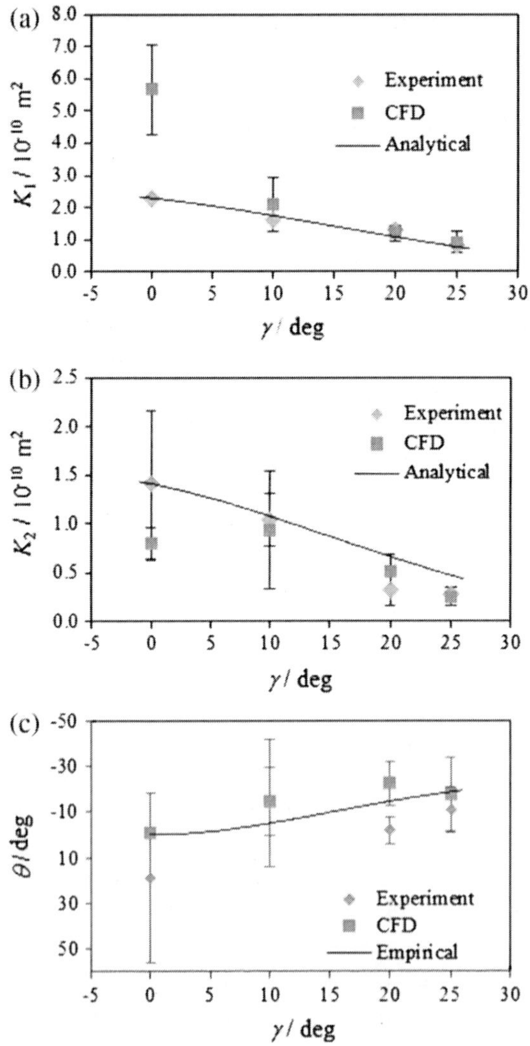

Figure 11. Three layers of plain weave fabric: predicted permeability (K_1, K_2 and ϑ) from CFD analysis as a function of the shear angle, together with experimental data and analytical/empirical predictions. (For interpretation of the references to colour in this figure legend, the reader is referred to the web version of this article.)

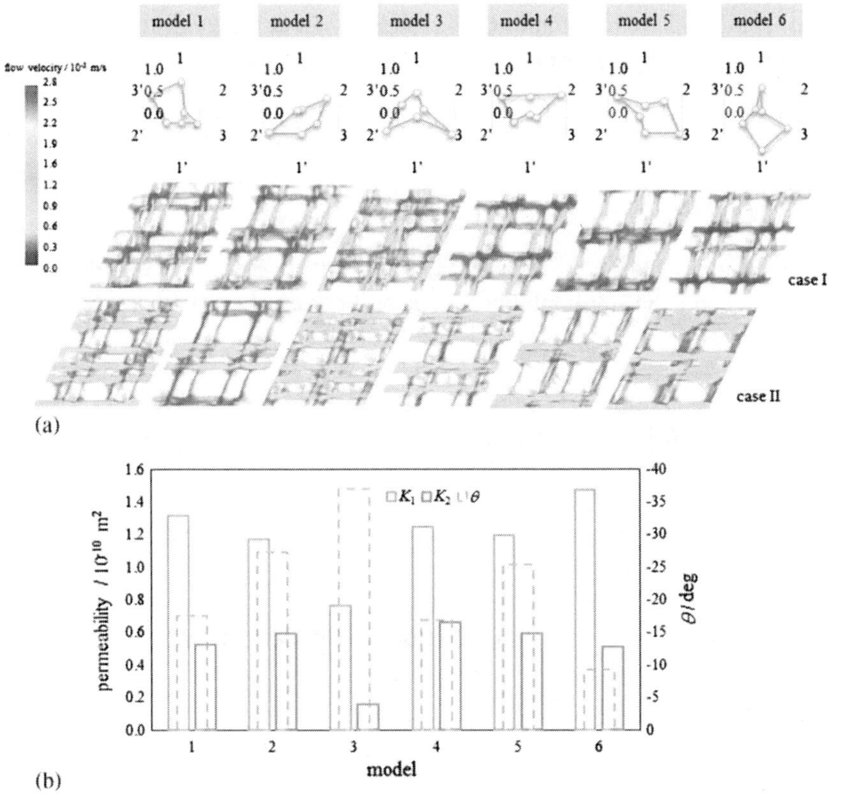

Figure 12. (a) Top: radar charts of random in-plane shift for the 3-layer unit cell models; the radial axis (0–1) is the normalised shift over the unit cell dimensions in warp and weft directions, whilst each spoke represents each of 3 fabric layers being shifted in warp (1, 2, 3) and weft direction (1′, 2′, 3′); (a) bottom: resulting distributions of flow channels in the sheared plain weave fabric lay-up (γ = 25°) due to the random shift, with predicted velocity field under the pressure boundary conditions case I and II as described in Section3; (b): corresponding permeability predictions (K_1, K_2 and ϑ) for the six stochastic models. (For interpretation of the references to colour in this figure legend, the reader is referred to the web version of this article.)

ACKNOWLEDGEMENTS

This work was supported by the Engineering and Physical Sciences Research Council [grant number: EP/IO33513/1], through the EPSRC Centre for

Innovative Manufacturing in Composites. X-ray μ-CT was conducted at the Hounsfield Facility, School of Biosciences, University of Nottingham.

REFERENCES

1. Aranda S, Berg DC, Dickert M, Drechsel M, Ziegmann G. Influence of shear on the permeability tensor and compaction behaviour of a non-crimp fabric. Compos Part B-Eng 2014;65:158–63.

2. Smith P, Rudd CD, Long AC. The effect of shear deformation on the processing and mechanical properties of aligned reinforcements. Compos Sci Technol 1997;57(3):327–44.

3. Heardman E, Lekakou C, Bader MG. In-plane permeability of sheared fabrics. Compos A Appl Sci Manuf 2001;32(7):933–40.

4. Louis M, Huber U. Investigation of shearing effects on the permeability of woven fabrics and implementation into LCM simulation. Compos Sci Technol 2003;63(14):2081–8.

5. Endruweit A, Ermanni P. The in-plane permeability of sheared textiles. Experimental observations and a predictive conversion model. Compos A Appl Sci Manuf 2004;35(4):439–51.

6. Verleye B, Croce R, Griebel M, Klitz M, Lomov S, Morren G, et al. Permeability of textile reinforcements: Simulation, influence of shear and validation. Compos Sci Technol 2008;68(13):2804–10.

7. Merotte J, Simacek P, Advani SG. Flow analysis during compression of partially impregnated fiber preform under controlled force. Compos Sci Technol 2010;70(5):725–33.

8. Endruweit A, Long AC. Analysis of compressibility and permeability of selected 3D woven reinforcements. J Compos Mater 2010; 44(24): 2833–62.

9. Endruweit A, Matthys KS, Peiro J, Long AC. Effect of differential compression on in-plane permeability tensor of heterogeneous multi-layer carbon fibre preforms. Plast, Rubber Compos 2009;38(1):1–9.

10. Grujicic M, Chittajallu KM, Walsh S. Effect of shear, compaction and nesting on permeability of the orthogonal plain-weave fabric preforms. Mater Chem Phys 2004;86(2–3):358–69.

11. Lomov SV, Verpoest I, Peeters T, Roose D, Zako M. Nesting in textile laminates: geometrical modelling of the laminate. Compos Sci Technol 2003;63(7):993–1007.

12. Hoes K, Dinescu D, Sol H, Parnas RS, Lomov S. Study of nesting induced scatter of permeability values in layered reinforcement fabrics. Compos A Appl Sci Manuf 2004;35(12):1407–18.

13. Endruweit A, Zeng X, Long AC. Effect of specimen history on structure and inplane permeability of woven fabrics. J Compos Mater 2014. 0021998314536070.

14. Arbter R, Beraud J, Binetruy C, Bizet L, Bréard J, Comas-Cardona S, et al. Experimental determination of the permeability of textiles: a benchmark exercise. Compos A Appl Sci Manuf 2011;42(9):1157–68.

15. Vernet N, Ruiz E, Advani S, Alms JB, Aubert M, Barburski M, et al. Experimental determination of the permeability of engineering textiles: Benchmark II. Compos Part a-Appl S 2014;61:172–84.

16. Nedanov PB, Advani SG. Numerical computation of the fiber preform permeability tensor by the homogenization method. Polym Compos 2002;23(5):758–70.

17. Chen Z-R, Lin Ye, Meng Lu. Permeability predictions for woven fabric preforms. J Compos Mater 2010;44(13):1569–86.

18. Song Y. Prediction of permeability tensor for three dimensional circular braided preform by applying a finite volume method to a unit cell. Compos Sci Technol 2004;64(10–11):1629–36.

19. Loix F, Badel P, Orgéas L, Geindreau C, Boisse P. Woven fabric permeability: from textile deformation to fluid flow mesoscale simulations. Compos Sci Technol 2008;68(7–8):1624–30.

20. Green SI, Wang Z, Waung T, Vakil A. Simulation of the flow through woven fabrics. Comput Fluids 2008;37(9):1148–56.

21. Liu HL, Hwang WR. Permeability prediction of fibrous porous media with complex 3D architectures. Compos A Appl Sci Manuf 2012;43(11):2030–8.

22. Lin H, Zeng X, Sherburn M, Long AC, Clifford MJ. Automated geometric modelling of textile structures. Text Res J 2012;82(16):1689–702.

23. Sherburn M. Geometric and mechanical modelling of textiles [PhD Thesis]: University of Nottingham; 2007.

24. McKay MD, Beckman RJ, Conover WJ. Comparison of three methods for selecting values of input variables in the analysis of output from a computer code. Technometrics 1979;21(2):239–45.

25. Olsson AMJ, Sandberg GE. Latin hypercube sampling for stochastic finite element analysis. J Eng Mech-Asce 2002;128(1):121–5.

26. Jiang D, Li Y, Fei Q, Wu S. Prediction of uncertain elastic parameters of a braided composite. Compos Struct 2015(0).

27. Gebart BR. Permeability of unidirectional reinforcements for Rtm. J Compos Mater 1992;26(8):1100–33.

28. Grouve WJ, Akkerman R. An idealised BC for the meso scale analysis of textile impregnation. The 9th international conference on flow processes in composite materials. Montreal 2008.

29. Advani SG, Sozer EM. Process modeling in composites manufacturing. CRC Press; 2002.

30. Vanaerschot A, Cox BN, Lomov SV, Vandepitte D. Stochastic framework for quantifying the geometrical variability of laminated textile composites using micro-computed tomography. Compos A Appl Sci Manuf 2013;44:122–31.

31. Gan J, Bickerton S, Battley M. Quantifying variability within glass fibre reinforcements using an automated optical method. Compos A Appl Sci Manuf 2012;43(8):1169–76.

32. Endruweit A, McGregor P, Long AC, Johnson MS. Influence of the fabric architecture on the variations in experimentally determined in-plane permeability values. Compos Sci Technol 2006;66(11–12):1778–92.

33. Long AC, Rudd CD, Blagdon M, Smith P. Characterizing the processing and performance of aligned reinforcements during preform manufacture. Compos Part A: Appl Sci Manuf 1996;27(4):247–53.

Chapter 11

Mechanical modelling of 3D woven composites considering realistic unit cell geometry

S.D. Green[a,*], M.Y. Matveev[b], A.C. Long[b], D. Ivanov[a], S.R. Hallett[a]

[a] Advanced Composites Centre for Innovation and Science (ACCIS), Department of Aerospace Engineering, University of Bristol, Bristol BS8 1TR, UK

[b] Polymer Composites Research Group, Division of Materials, Mechanics and Structures, University of Nottingham, Nottingham NG7 2RD, UK

ABSTRACT

Modelling the mechanical performance of textile composites is typically based on idealised unit cell geometry. However, 3D woven composites feature more complex textile architecture then 2D woven materials, and in reality nominally straight warp and weft yarns can also possess significant waviness. For such textiles, idealising yarns as straight entities becomes an oversimplification. In this study, the voxel method and a continuum damage model are used in a finite element analysis to compute stress–strain curves for an orthogonal 3D woven composite under tensile loading. The main goal of this study was to compare results produced using idealised geometry with realistic geometry obtained from detailed simulation of the preform during weaving and compaction. Significant variation in predictions was obtained using the different geometrical models. The idealised model lead to an overestimation of stiffness and strength compared to experiment due to the neglecting of yarn waviness, whereas the simulated geometry models produced more conservative results closer to experiment.

KEYWORDS

3D woven composite; Unit cell; Finite element analysis (FEA); Virtual test; Voxel

1. INTRODUCTION

3D woven composites can provide a potential solution to the fundamental limitations of traditional laminated composites; delamination and labour intensive manufacture. The addition of binder yarns provides through-thickness reinforcement leading to greatly enhanced interlaminar properties as well as binding the fabric to enable near-net-shape preforms to be woven and handled. However, despite these advantages, the use of 3D woven composites has been largely limited to niche applications. One of the key reasons for this is the lack of predictive numerical tools, which limits their ability to be used at the early stages of design.

Mechanical performance modelling of textile composites typically begins with the definition of textile unit cell geometry using a specialist pre-processor such as TexGen[1] and [2] or WiseTex [3] and [4]. Such software can produce an idealised representation of a textile, sufficient for the modelling of many types of composites with 2D reinforcement. However, some 3D woven textiles present a significant challenge to model due to their inherent complexity. While TexGen is capable of modelling such textiles, the idealised geometries that it produces can neglect realistic features such as yarn waviness and yarn pinching which play a significant role in determining their resulting properties, especially strength [5] and [6]. Previous work [7] used a finite element (FE) model, based on the digital element method [8] in order to predict the deformations of an orthogonal 3D woven fabric at the unit cell level. Each yarn was represented as a bundle of 61 chains of beam elements, with contact between chains within a yarn defining yarn cross-sectional shape and contact between each yarn assembly defining the yarn paths. Tensioning of the binder yarns was used to simulate the weaving process prior to moulding through compaction between rigid plates. Throughout this procedure the unit cell was subjected to periodic boundary conditions such that the resulting geometry was periodic.

Conventional FE models of 2D woven textile composites are typically meshed with tetrahedral elements, with the matrix domain designated as the inverse of the yarn volume within the global unit cell domain (see for example [9]). Nodes of matrix and yarn elements at the interface between the two different material phases must be shared to form a conformal mesh. However, realistic models of complex 3D woven architectures will likely feature degenerated resin regions and small yarn interpenetrations meaning that such meshing practices are no longer feasible, especially when a periodic mesh is required. Other methods have been proposed to model such textile composites including the domain superposition technique (DST) [10] and embedded element technique [11]. In these methods, the matrix domain is assigned as the global domain leaving the yarns to be meshed independently. A coupling technique between the domains ensures continuity of displacements while properties of the yarns are reduced to account for the superposition. The independent mesh method (IMM) [12] is a similar technique where any matrix elements lying fully within yarn domains are excluded from the analysis. Matrix elements which lie on the boundary are then subjected to a refinement scheme. The extended finite element method (XFEM), originally developed for modelling discontinuities due to cracks, has also been applied to heterogeneous materials with discontinuities at the material interface [13], including textile composites [14]. The global domain is meshed using regular hexahedral elements and the nodes of elements in the interface region are enriched with extra degrees of freedom to account for the strain field jump. However, it is still required that the geometry is free from interpenetrations. In this study, a technique based on voxels [15]; meaning a 3D pixel, is used. The global domain is meshed in a similar process to XFEM and elements are assigned to either the matrix or yarn domain depending on the location of the element centroid. The voxel method has been shown to be adequate for mechanical modelling of 2D textile composites [16].

2. TEXTILE GEOMETRY

The accuracy of mechanical modelling of textile composites for strength prediction is dependent on the modelling technique, material models and assumed textile geometry. A previous project considering failure modelling of 3D woven composites [17] concluded that a significant limitation was with regard to the idealised geometries being used. For textiles with complex internal architecture, there was no further merit in pursuing relatively small improvements with the failure models whilst significant discrepancies existed between considered and real geometries. A schematic representation of an orthogonal 3D woven fabric is shown in Fig. 1. The fabric has two sets of binder yarns, each arranged in a 5 harness satin style. One set of these binder yarns float on the upper surface of the fabric while the other float on the bottom surface. The original idealised TexGen unit cell model of this fabric used in the aforementioned study is shown in Fig. 2(a). In this geometrical model, the in-plane warp and weft yarns are completely straight with zero crimp, waviness or cross-sectional variation. Binder yarns follow a path with straight horizontal or vertical sections, constant cross-section and extend beyond the surface of the fabric, resulting in a resin layer on the surfaces of the composite. None of these features are representative in the real moulded composite.

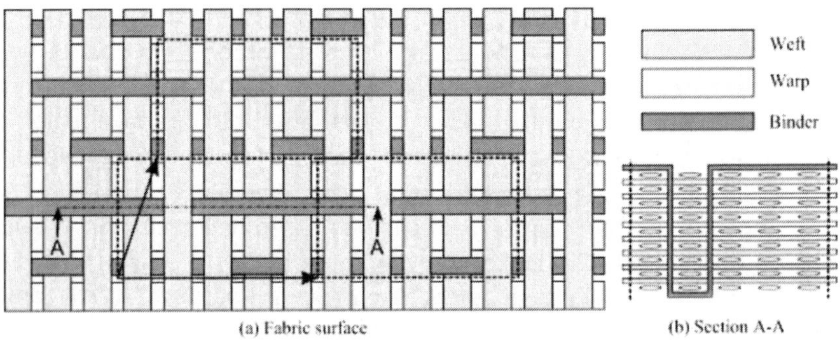

Figure 1. Schematic representation of orthogonal 3D woven fabric showing; (a) unit cell with tessellation and (b) fabric cross-section.

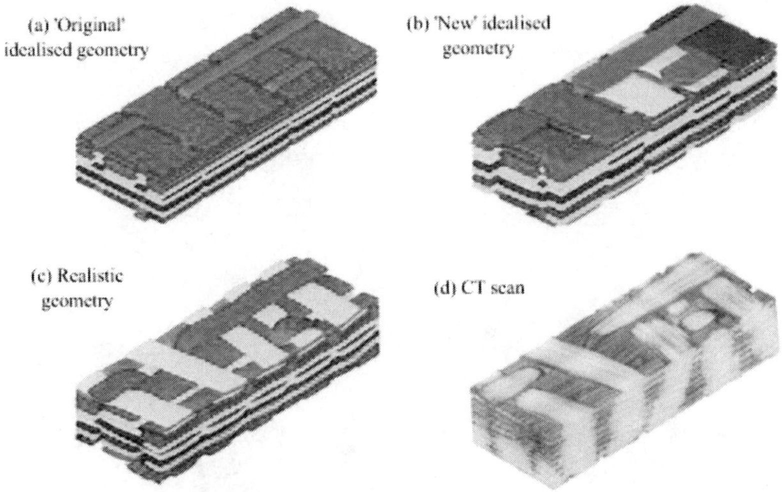

Figure 2. Various TexGen unit cell models of 3D woven fabric compared to CT scan, at 52% o-FVF.

However, there have been recent developments to TexGen aimed specifically at improving the quality of 3D woven textile geometry. Despite this, it was not possible to generate a model at the level of compaction consistent with the real infused composite (58.5% overall fibre volume fraction (o-FVF)) using the latest version of TexGen (3.5.3).Fig. 3 shows an attempt at generating such a model. Fig. 3(a) shows excessive squeezing of binder yarns at the surface while Fig. 3(b) shows penetrations of binder and weft yarns. A valid model at the target o-FVF could only be achieved by shrinking yarn cross-sections, thus requiring excessive intra-yarn fibre volume fraction (iy-FVF). As such, the new idealised geometrical model used in this study has a lower o-FVF of 52.0%, similar to the original idealised model, and is shown in Fig. 2(b). This model features binder yarns which are flush with the fabric surface, achieved using binder yarns which are near circular in the vertical sections but flatten considerably on surface while inducing crimp in the surface weft yarns. However, elsewhere the in-plane yarns remain straight with constant cross-section.

A numerical model has been developed and used to predict the deformations of the 3D woven fabric considered in this study during weaving and subsequent compaction in a mould tool [7]. This was conducted in order to produce accurate geometry for unit cell analysis of textile composites. Whilst these deformation models can produce geometry at levels of compaction as high as the real fabric, for comparative purposes, this simulated geometry TexGen model is also shown at 52.0% o-FVF in Fig. 2(c), and is visually considerably different to the idealised models. It features significant in-plane and out-of-plane waviness in warp and weft yarns and variation of cross-sections in all yarns.

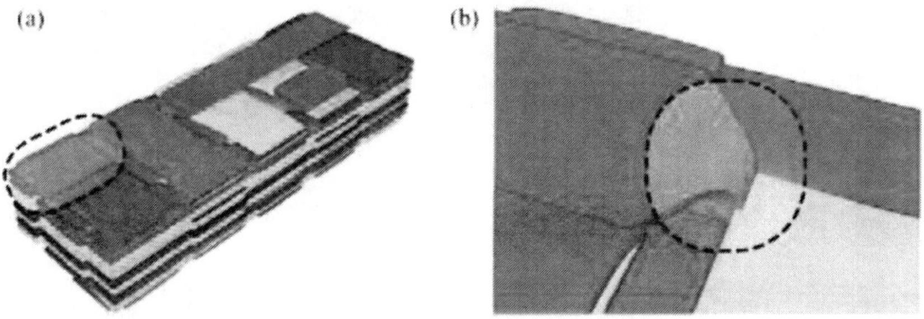

Figure 3. Idealised TexGen model at 58.5% o-FVF showing; (a) excessive yarn flattening, (b) example of yarn interpenetrations.

The average waviness levels in each of the geometrical models are listed in Table 1. This was determined by splitting the yarn path into short segments and calculating the angle of each segment from the nominal path both in-plane (ϑ_{IP}) and out-of-plane (ϑ_{OOP}), as shown in Fig. 4 and Eq. (1). The overall waviness ($\vartheta_{overall}$) of each segment was assessed by considering a 3D vector, *a*, defined by ϑ_{IP} and ϑ_{OOP}. $\vartheta_{overall}$ was calculated as the angle of *a* with another vector, *b*, defining nominal yarn path, as calculated in Eq. (2). The final waviness value is the average of this value over the total length of the yarn.

$$\theta_{IP} = \tan^{-1}\left(\frac{z_1 - z_0}{x_1 - x_0}\right), \quad \theta_{OOP} = \tan^{-1}\left(\frac{y_1 - y_0}{x_1 - x_0}\right) \tag{1}$$

$$\theta_{overall} = \cos^{-1}\frac{a \cdot b}{|a||b|} \tag{2}$$

$$\text{where}: \begin{array}{l} a_{warp} = (1, \tan\theta_{IP}, \tan\theta_{OOP}) \\ b_{warp} = (1, 0, 0) \end{array}$$

Table 1. Average waviness values in each of the models.

	Ideal (original)	Ideal (new)	Simulated (52.0% o-FVF)	Simulated (58.5% o-FVF)
Warp	0°	0°	3.14°	4.81°
Weft	0°	0.19°	3.91°	4.11°

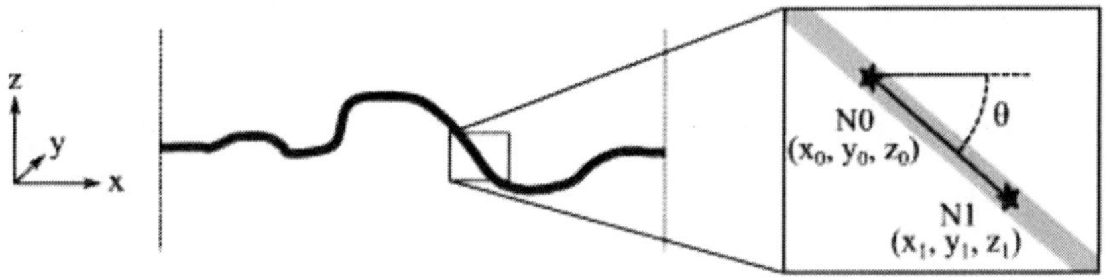

Figure 4. Path of a warp yarn, showing method to define waviness.

The output of the deformation model has been validated against computed tomography (CT) scans of the dry fabric subjected to in-situ compaction across a range of volume fractions in [7], with excellent qualitative correlation. A CT image of a unit cell is presented in Fig. 2(d) where it can clearly be seen that the simulated geometry model represents the architecture of the real fabric much more accurately than the idealised models, but unfortunately experimental data for the waviness of the fabric at the high level of compaction considered in this paper is not available.

2.1. Conversion of realistic geometry from deformation models

The output of the deformation models, with yarns represented as a bundle of beam elements, is not immediately useful for meso-scale mechanical performance analysis where yarns require definition as solid geometry. Conversion of the deformation model geometry was performed using Python scripting to create the simulated geometry TexGen model [18] shown in Fig. 2(c), from which FE models can be generated. Building a unit cell textile model in TexGen requires definition of three attributes; yarn paths, yarn cross-sections as well as yarn repeats with a domain. For detailed information regarding the theory of geometrical modelling of textiles as applied in TexGen, refer to [1].

The path of a yarn in TexGen is represented by a one-dimensional line defined in three-dimensional space. The script was used to march along each yarn, stopping at regularly spaced intervals to calculate the yarn centroid and attribute these as master nodes which were joined with a periodic spline (Fig. 5). Yarn cross-sections are defined in TexGen as two-dimensional shapes in a plane perpendicular to the yarn tangent. Planes were specified at each master node to account for the variation in cross-sectional shape along the yarn path and the intersection points of each beam element within the yarn was mapped to these planes. Next, cross-sections were defined based on these mapped points, using the procedure outlined below to generate polygons. These polygons are sampled by a number of equally spaced slave nodes for graphical rendering and meshing.

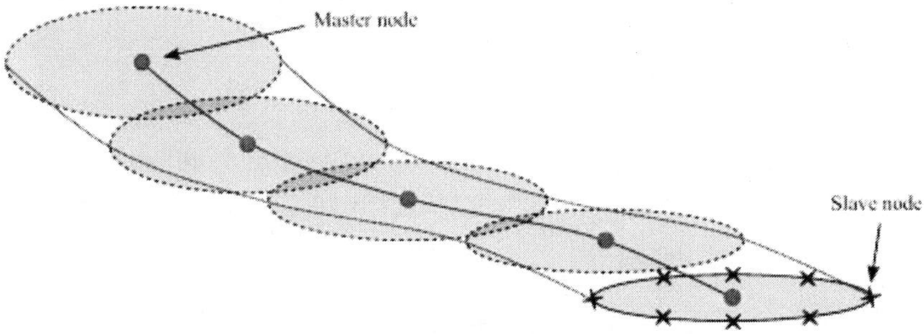

Figure 5. Yarn definition in TexGen.

Convex hull algorithms are widely used in computational geometry to define the smallest convex polygon around a set of points e.g. [19] and [20]. This can be visualised using an analogy in which the points are represented by pins in a board. By tying a piece of string to a starting pin, the convex hull is the shape formed by wrapping the string around the set of the pins until it returns to the starting pin. However, the shape of a yarn cross-section is not always convex and applying the convex hull would give an inaccurate representation of the yarn in such cases. Therefore a modified convex hull algorithm was implemented, beginning by defining the first point in the polygon as the node with the largest x co-ordinate (in an x–y plane), then proceeding anti-clockwise around the section. Using the previous analogy, a string of finite length, r, is employed to define a search radius for the subsequent selection which is the point creating a line with the smallest external angle from the previous polygon section. A closed polygon is formed once the initial node is reselected and is subsequently expanded by the radius of the beams, $d/2$. The search radius, r, was defined as $r = xd$, where the parameter x can be used to adjust the level of smoothing of the cross-section. A value of $x = 4$ was found to produce good results and is compared with the convex hull solution for two extreme examples of yarn cross sections generated by the deformation models in Fig. 6. While the convex hull produces good results for the thick yarn, its representation of the thinner yarn is poor since the large concave regions are smoothed over. The modified algorithm, however, produces good results in both situations.

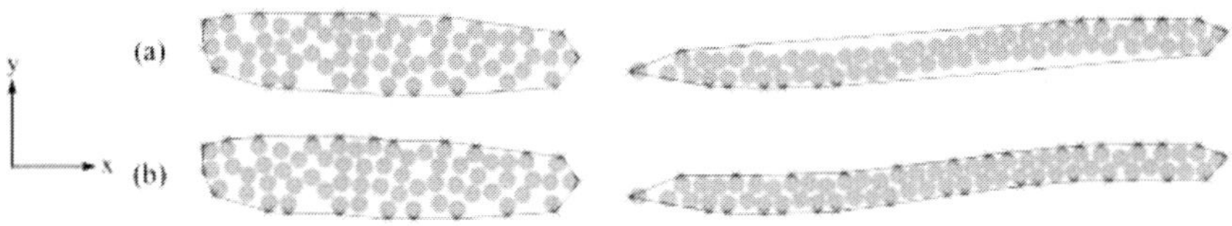

Figure 6. Illustration of; (a) convex hull ($x = \infty$) and (b) modified convex hull algorithms ($x = 4$) as applied to two different yarn shapes as generated by deformation model.

Since the output of the unit cell deformation models is periodic, specification of yarn repeat vectors corresponding to the tessellation of the unit cell can be utilised in TexGen to define an infinite fabric which is then trimmed to a finite domain of the unit cell.

3. MODELLING METHODOLOGY

In this study, tensile loading in warp and weft directions will be considered for the following four scenarios:

A. New idealised geometry model (52.0% o-FVF)

B. Simulated geometry model (52.0% o-FVF)

C. Simulated geometry model (58.5% o-FVF)

D. Experimental data (58.5% o-FVF)

The quoted o-FVFs for A–C refer to the TexGen geometrical models. The voxel models generated from these TexGen models show negligible difference from these values, since any errors in yarn volume fraction (YVF) resulting from the voxel discretisation are offset by adjustment of the iy-FVF. The results of A and B are used to assess the effect of the different geometries on mechanical performance prediction, since the two models can be directly compared at the same o-FVF. Further insight can be gained regarding the effect of compaction from B to C which can only be achieved by using models of the realistic geometry. The accuracy of each model as compared to experiment will also be assessed.

3.1. Mesh and boundary conditions

Voxel meshes can be automatically generated in TexGen with each voxel being designated to either a yarn or the matrix depending on the location of its centroid. This means the technique is tolerant of small yarn interpenetrations in the geometrical models which can be introduced during the conversion process from deformation models. Unit cell models of size 27.8 × 9.9 × 5.3 mm were used with cubic shaped voxel elements providing an element length of 0.12 mm. Using this technique, it is trivial to generate periodic meshes of complex geometries thus reducing pre-processing effort and facilitating automation. It also avoids the need to specify the fictitious resin gaps between yarns often resorted to in conventional modelling of textile composites to produce adequate quality elements (see for example [21]). Material orientations were assigned to each voxel element using the procedure applied in TexGen which is outlined in [1]. Periodic boundary conditions for the unit cell with stagger in tessellation shown in Fig. 1(a) were applied as derived in [22]. Since the unit cell represents the full thickness of the material, the upper and lower surfaces were unconstrained.

3.2. Material properties

In each of the models the yarn cross-sectional area and hence iy-FVF varied in segments along yarn path. This variation was from 47–78% iy-FVF in the idealised model, though most of this variation occurs in the binder yarns. The simulated geometry models exhibited a high diversity of yarn shapes and dimensions in all yarns with iy-FVF ranging from 36–87%, where high iy-FVFs occurred in regions of yarn to yarn contact at cross-over points where load is transmitted through fabric thickness during weaving and compaction. Whilst the iy-FVF remains below the maximum hexagonal array fibre packing limit of ~91%, since this cannot be violated in such multi filament based deformation models, the high iy-FVFs suggested in certain localised regions is exaggerated. In practice it is difficult to exceed iy-FVFs greater than 70–75% (see for example [23], [24] and [25]) as unlike the model, fibres in a real yarn are not perfectly aligned and have some level of entanglement.

The resin pockets were assigned the properties of the pure isotropic epoxy resin while the yarns were treated as transversely isotropic homogenous material similar to a unidirectional lamina. Properties were applied to each element within the yarn depending on the local iy-FVF, in order to accurately represent the distribution of properties along the yarns. Fibre and matrix properties are shown alongside estimated yarn properties at a nominal iy-FVF of 70% in Table 2. Properties of the carbon fibres and epoxy resin were taken from manufacturer's data, though some fibre properties were taken from the literature, with averaged experimentally derived values used where available.

Yarn elastic constants were calculated from constituents using a micromechanical FE model with hexagonal fibre packing. Strength properties were estimated using a simple analytical model, called the generalised chess model, first used in [26]. Fig. 7(a) shows the micro scale unit cell of a unidirectional fibrous composite with regular square packing. This unit cell has two planes of symmetry and thus the considered region can be reduced to a quarter of this unit cell. Fig. 7(b) shows the quarter unit cell approximated by a square fibre grid consisting of four blocks. Blocks 1, 2 and 4 represent matrix material with block 3 representing the fibre. A set of conditions including the prescribed stress/strain field, continuity of traction on the inter-block boundaries and symmetry form a system of equations relative to the stress in each block. In essence, the obtained representation is similar to the discretisation of the unit cell by four finite elements.

These equations along with the Hooke's law are sufficient to identify the stresses in each block, which are assumed to be uniform. Once the stresses are determined, strength can be estimated. The fibre dominated longitudinal strength is calculated by maximum stress criterion and resin dominated strength properties are calculated using the following criteria:

$$\max\left(\frac{p_{mx}}{p_{cr}}, \frac{q_{mx}}{q_{cr}}\right) \geqslant 1 \qquad (3)$$

Table 2. Table of material properties for constituents and yarn (at nominal 70% iy-FVF).

	E_{11} (GPa)	E_{22} (GPa)	$\nu_{12} = \nu_{23}$	ν_{23}	$G_{12} = G_{13}$ (GPa)	G_{23} (GPa)	S^{s}_{11} (MPa)	S^{s}_{22} (MPa)	$S^{s}_{12} = S^{s}_{13}$ (MPa)	S^{s}_{23} (MPa)
Carbon fibre [40]	238	13 [35,36]	0.20 [41]	0.25 [41]	13 [42,43]	6 [41]	4620[a]/3825[b]	–	–	–
Epoxy resin MVR444 [44]	3.1	3.1	0.35	0.35	1.2	1.2	77.6	77.6	61.5	61.5
Yarn (70% iy-VF)	167	8.1	0.24	0.37	4.5	3.0	3234[a]/2678[b]	36.4	53.8	61.5

[a] Warp/weft yarns (HTS40 F13 12k).
[b] Binder yarns (HTA40 E13 6k).

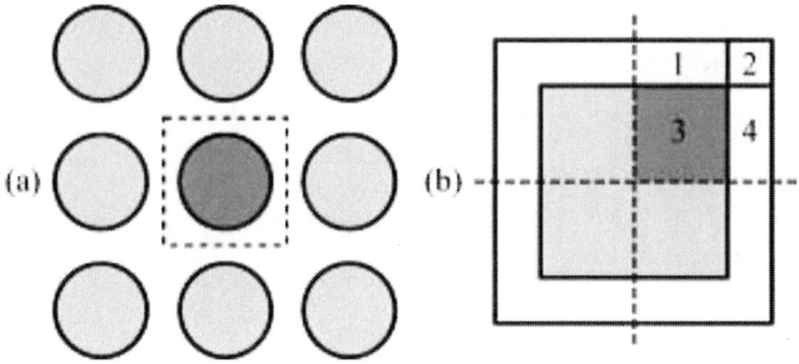

Figure 7. Illustration of generalised chess model showing; (a) micro scale unit cell of unidirectional composite with square packing, (b) chess model discretisation of unit cell, reduced to a quarter by symmetry. (In each image, shaded and blank regions represent fibres and matrix respectively).

where P_{mx} and P_{cr} are the maximum hydrostatic tension in the matrix (crucial for failure of brittle epoxies [27] and [28]) and its critical value. In an elementary uniaxial tensile test the epoxy matrices fail in a brittle fashion and hence $p = p_{cr} = \frac{S^t}{3}$ at failure (one third of the apparent uniaxial strength S^t). q_{mx} and q_{cr} are the Mises stress and the correspondent critical value characterising the probability of a ductile failure. In pure shear the epoxy yields as an ideal plastic material before failing, hence q_{cr} can be taken as the yield stress.

The model was used to predict strength under three uniaxial conditions; transverse tension, in-plane shear, out-of-plane shear. The prediction of the transverse tensile strength follows the anticipated trend; strength drops when iy-FVF is increased from 0% to 40% and remains nearly constant as it further increases. The transverse tension and shear strength values are in good agreement with the available data, e.g. [29].

3.3. Material damage models

Carbon fibres often exhibit non-linear tensile behaviour, e.g. at an applied strain of 1% the Young's modulus may increase by 20–25% of its initial value [30]. Recent studies showed this phenomenon (known as the 'non-Hookean' effect) to be partly responsible for the non-linear behaviour of 3D woven composites [31]. However, due to the lack of data for the particular carbon fibres used here, the behaviour of yarns was assumed to be linear prior to damage initiation. The matrix was treated similarly. After damage initiation a continuum damage mechanics (CDM) model [32] was applied both to yarns and matrix. It assumes that damage in a material can be considered as a reduction of stiffness properties over a region of material, a single finite element in this case.

Matrix damage was assessed by the pressure dependent modified von Mises criterion listed in Eq. (4)[33]. Damage initiation was assumed when the damage parameter $D_m$exceeded the value of 1.0.

$$D_m = \frac{S_m^c - S_m^t}{S_m^c S_m^t}(\sigma_1 + \sigma_2 + \sigma_3) + \frac{1}{2S_m^c S_m^t}\left[(\sigma_1 - \sigma_2)^2 + (\sigma_2 - \sigma_3)^2 + (\sigma_1 - \sigma_3)^2\right] \quad (4)$$

where σ_i with i = 1, 2, 3 are the principal stresses and S_m^j with $j = t, c$ are the matrix tensile and compressive strengths.

Three modes of failure were considered for the yarns: longitudinal tension, transverse tension and shear. The damage state of the material was defined by a set of three parameters D_i listed in Eqs. (5), (6) and (7) with indices i = 1, 2, 3 representing longitudinal, shear and transverse failure modes respectively.

$$D_1 = \max\left(\frac{\sigma_{11}}{S_{11}^t}, \frac{-\sigma_{11}}{S_{11}^c}\right) \quad (5)$$

$$D_2 = \frac{\sqrt{\sigma_{12}^2 + \sigma_{13}^2}}{S_{12}} \quad (6)$$

$$D_3 = \max\left(\frac{\max(\sigma_2, \sigma_3)}{S_{22}^t}, -\frac{\min(\sigma_2, \sigma_3)}{S_{22}^c}\right) \quad (7)$$

where; σ_{ij} are components of stress tensor, S_{ij}^k are strengths of yarn with indices i, j = 1, 2 corresponding to the directions while $k = t, c$ stands for tension and compression.

The moduli of the damaged material were described by Eqs. (8), (9) and (10) for the yarns and Eq. (11) for the matrix:

$$E_{11} = \begin{cases} E_{11}^{0}, & D_1 \leqslant 1 \\ 0.001E_{111}^{0}, & D_1 > 1 \end{cases} \tag{8}$$

$$E_{22} = E_{33} = E_{22}^{0}\max(0.001, min(P(D_2), P(D_3))) \tag{9}$$

$$G_{12} = G_{13} = G_{12}^{0}\max(0.001, min(P(D_2), P(D_3))) \tag{10}$$

$$E_{m} = E_{m}^{0}\max(0.001, P(D_m)) \tag{11}$$

where stiffness properties with a superscript 0 refer to the undamaged value and $P(D_i)$ is a penalty function defined as:

$$P(D_i) = \left(1 - \frac{1}{\exp(-c_1 D_i + c_2)}\right) \tag{12}$$

It can be seen that damage in the longitudinal direction causes abrupt reduction of longitudinal properties whereas other stiffness properties are subject to an exponential degradation. Poisson's ratios ν_{12}, ν_{23} and ν_{13} remain intact, while the corresponding Poisson's ratios in other directions (i.e. ν_{21}, ν_{32} and ν_{31}) were adjusted to keep the stiffness matrix of an element symmetric. Parameters c_1 and c_2 in the penalty function are empirical constants where their ratio determines how fast damage will propagate through an element and defines the final failure of the element (Fig. 8). This will have some mesh dependence, with coarse meshes requiring a higher ratio to prevent the dissipated energy increasing. The ratio c_2/c_1 was initially chosen from a model of a 2D woven unit cell [16] with a more refined mesh. Based on the initial results of loading in the warp direction an improved value of 4.0 was selected. The influence of mesh size whilst keeping this parameter constant has been investigated and results are discussed in Section 4.

4. RESULTS AND DISCUSSION

Voxel meshes used for the idealised and simulated geometry models are shown in Fig. 9. It was relatively simple to represent the straight yarns of the idealised model with voxels, but the complex geometry of the simulated geometry models proved more challenging with angled or curved yarns having stepped or jagged surfaces. The quality of the voxel representation of geometry improves with refinement, however the large unit cell size of this fabric placed a practical limit of the level of refinement possible. The baseline models had nearly 800,000 elements and took 5 days to run on a high performance computing system using 32 CPUs.

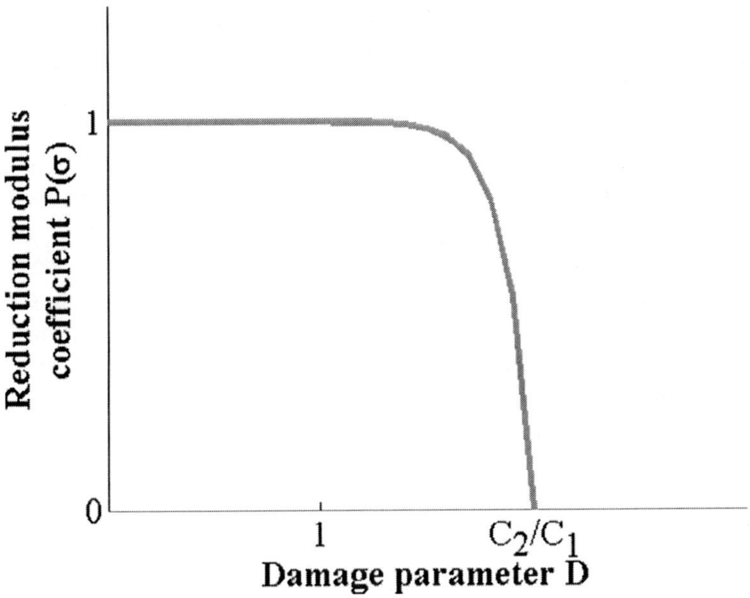

Figure 8. Stiffness reduction penalty function.

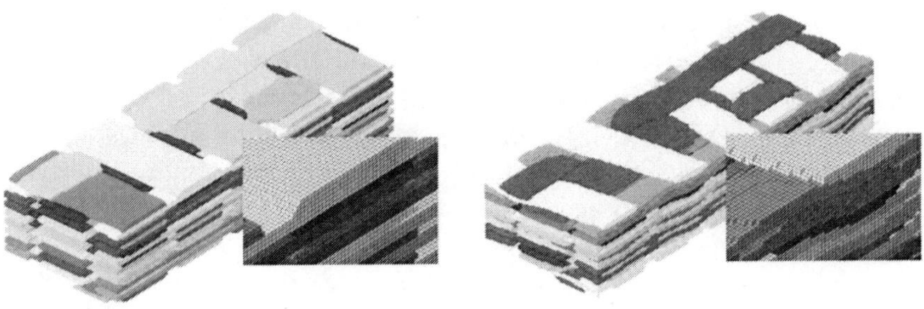

Figure 9. Voxel meshes for; (a) idealised model, (b) simulated geometry model.

Young's moduli and ultimate tensile strength values are presented for each of the models compared to experiment in Table 3. Stress–strain curves for each data set are shown for the warp and weft directions in Figs. 10 and 11 respectively. Experimental curves [34]shown are for a representative test and model results using the low compaction geometry have been normalised by o-FVF to aid comparison.

Non-linear behaviour of a 3D woven composite was shown in [31] to be the result of a combination of three phenomena: fibre non-linearity, inelastic yarn straightening and damage development in the composite. The latter of these two phenomena are accounted for in the modelling here. It was also shown in [30] that the first two phenomena dominate at lower strains while damage development overcomes them at higher strains.

The experimental curves in Figs. 10 and 11 are bi-linear with a stiffness reduction kink at around 0.6% strain. It is believed that this non-linearity is predominantly due to damage development, more specifically transverse yarn, and also resin, cracks. These have been observed in an orthogonal 3D composite of similar weave style but fewer layers in interrupted tests (Fig. 12). Significant damage is also present in the models prior to final failure, see Fig. 13 showing 'cracks' of damaged material forming in the yarns perpendicular to the applied load.

Longitudinal strains in the yarns were assessed in one of the models. It was found that at an applied strain of 0.6% (where ~10% of transverse yarns material has sustained some damage) only a small fraction (<1%) of the yarns have strain values higher than 1%. Based on [31], this implies that overall increase of stiffness cannot be higher than 10–15%. It was therefore considered reasonable to neglect fibre non-linearity in the material model.

The result of the idealised model was an overestimation of stiffness in both directions by around 15%. The low compaction simulated geometry model predicted moduli very close to experiment with the high compaction model having slightly reduced values. The same trend is replicated with strength prediction for each model. The damage model used for the analysis is based on a simplified mechanical approach and thus cannot capture transverse failure with great accuracy. Stress distribution, damage onset and final strength are significantly influenced by the yarn crimp. Hence, the simulated geometry models produced a more accurate prediction of the strain onset of transverse yarn damage accumulation and the stress–strain curve shows the main trend of stiffness reduction. The final failure depends mainly on correct prediction of longitudinal stresses, although the level of damage present at the point of failure affects maximum stress which can be sustained.

Table 3. Comparison of modulus (*E*) and ultimate tensile strength (S^t) predicted by the models and from experiment.

	Ideal (52.0% o-FVF)[a]		Simulated (52.0% o-FVF)[a]		Simulated (58.5% o-FVF)		Experiment (58.5% o-FVF)	
	E (GPa)	S^t (MPa)	E (GPa)	S^t (MPa)	E (GPa)	S^t (MPa)	E (GPa)	S^t (MPa)
Warp	72.8 (64.7[a])	922 (820[a])	64.3 (56.5[a])	669 (615[a])	56.7	551	63.9 (0.73[b])	701 (35.0[b])
Weft	71.0 (63.1[a])	920 (818[a])	61.7 (54.6[a])	596 (530[a])	59.7	533	60.8 (0.80[b])	625 (40.9[b])

[a] Raw, non-normalised data.
[b] Standard deviation.

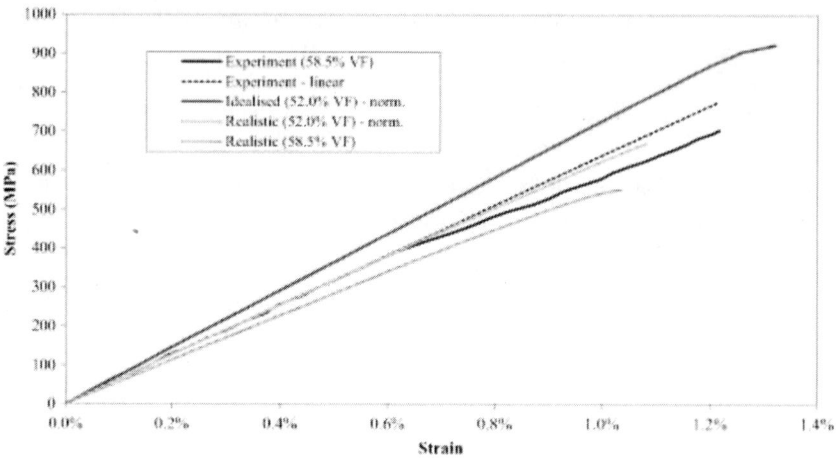

Figure 10. Graph showing stress strain curves of each model alongside experiment for tensile warp loading. Data from 52.0% o-FVF models is normalised.

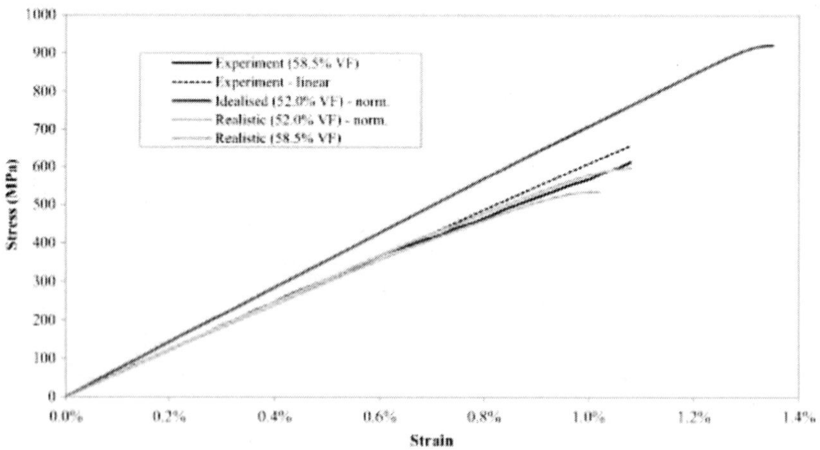

Figure 11. Graph showing stress strain curves of each model alongside experiment for tensile weft loading. Data from 52.0% o-FVF models is normalised.

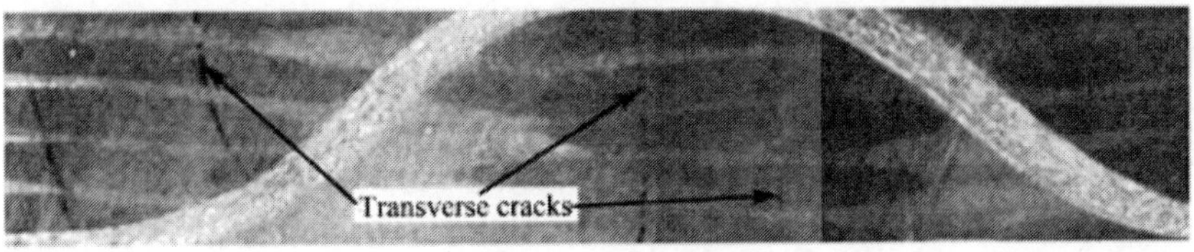

Figure 12. Micrograph of a thinner orthogonal 3D woven fabric of similar weave style showing transverse cracks in weft yarns and resin at 90% failure load in interrupted test [34].

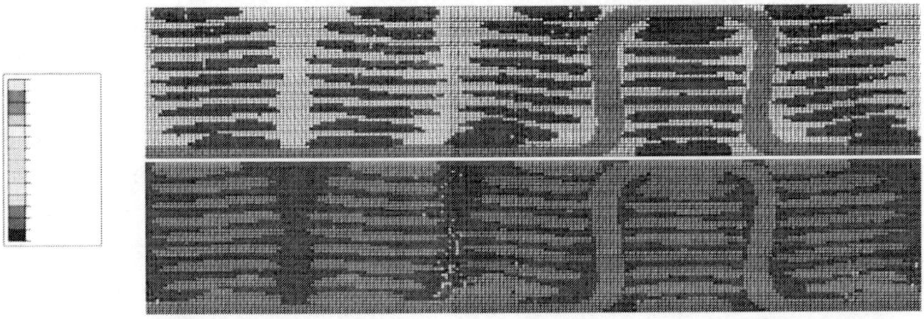

Figure 13. Damage plots in 58.5% o-FVF simulated geometry model at 1% strain showing; (a) transverse tensile damage in weft yarns, (b) resin damage, where 0 and 1 represent undamaged and fully degraded material respectively.

Comparison of the two 52% o-FVF models clearly demonstrates the effect of textile geometry on the resulting mechanical properties. Non-conservative results were produced by the idealised model because of the geometrical assumptions outlined in Section 2, most notably the lack of yarn waviness. Assessment of the two simulated geometry models, considering the normalised data for the low compaction model, shows a reduction in mechanical properties with compaction since the highly compacted model features greater waviness. In these cases, the relative reduction in tensile strength was over double the relative reduction in stiffness, illustrating that strength is more sensitive to geometrical defects than stiffness. This effect will be even more severe for compressive strength since the presence of waviness causes kink-band formation in yarns [5] and [6]. Even when considering the raw, un-normalised data of the low o-FVF simulated geometry model, the results suggest that the mechanical properties do not necessarily improve with compaction since the effect of greater o-FVF is counteracted by the reinforcement acting with lower efficiency. This indicates that there is likely to be an optimum moulded thickness for a given reinforcement type and desired properties with regards to a bias towards stiffness or strength. The use of deformation models in conjunction with failure models outlined in this paper could be used help determine this thickness for a given weave style, since the behaviour of 3D woven composites is very dependent on the specific reinforcement and results from one textile cannot be considered to carry over to a different textile.

One would expect the fully compacted simulated geometry model to produce the closest match with experiment, however predicted properties were slightly underestimated. One potential source of error was the applied boundary conditions. A unit cell with periodic boundary conditions behaves as if it is part of an infinite material, although due to the physically large unit cell size specimens for testing were only around one unit cell wide for weft loading and two for warp loading. Running unit cell models with no periodicity constraints in the lateral direction however showed negligible changes in stiffness, thus eliminating this as a possible cause. The results of these models are sensitive to material inputs and whilst every care was taken in determining these properties, it is possible that inaccuracies exist between the real material and properties used due to the reliance on manufacturer's and literature data. Fibre transverse modulus for example has a notable effect on overall stiffness and published values vary widely [35] and [36]. However this was not attributed as the main source of error.

As further investigation, and in order to assess the mesh sensitivity of the models, a convergence study was conducted, generating coarse and refined models by dividing and multiplying the baseline element dimensions by a factor of 1.5 giving element sizes of 0.18 and 0.08 mm respectively. This was done for weft loading in the fully compacted realistic model since there was more damage prior to final failure and this would indicate any differences between different meshes most clearly. The results are summarised inTable 4. The predicted modulus was similar for the coarse and baseline models, but increasing refinement showed a small increase in modulus. This trend of increasing stiffness with voxel refinement occurs for yarns with misalignment to the mesh direction since the voxel discretisation gives a stepped surface which is structurally less efficient. However, when considering straight yarns aligned to the voxel edges, voxel models produce an accurate stiffness prediction compared to a traditional

FE approach, even with low mesh refinement. High refinement of misaligned yarns will cause the prediction to converge to the correct numerical solution, though the level of refinement used in this paper is coarser than what is required to achieve full convergence. The implication of this is that the voxel method accurately modelled the idealised geometry but underestimated stiffness for the two more accurate simulated geometry models.

Table 4. Comparison of modulus (*E*) and ultimate tensile strength (S^t) predicted by the models in the convergence study.

Relative element length	Number of elements	E (GPa)	S^t (MPa)
0.66	231,000	60.2	585
1.0	787,500	59.7	533
1.5	2,627,250	62.3	562

The damage parameter c_2/c_1 was kept constant in each model such that the strength predictions in the convergence study illustrated the effect of mesh refinement on the damage model. Stiffness degradation due to damage could therefore be expected to retard with refinement because the energy dissipated reduces. However, the effect of refinement on strength did not follow this trend, where in fact there was no clear pattern with both coarse and refined models having a slightly higher strength than the baseline. Extraction of the force through a single weft yarn in the centre of the model at 0.9% strain showed a similar trend to the ultimate strength predictions. The variation in strength was largely due to the variation in voxel discretisation for the different mesh sizes causing variations in the local stress state which fluctuate with mesh refinement rather than converging on a single value. It can thus be concluded that for the range of meshes investigated here, the damage formulation itself is not highly sensitive to the mesh size, once an initial set of parameters is chosen, with voxel discretisation effects dominating. This is a limitation of the voxel method when applied to complex geometrical models, since the key advantage of the technique is the ability to mesh complex geometry. Several techniques could be implemented to improve the accuracy of voxel modelling at low refinement. These include a better representation of the yarn–matrix interface through local refinement [15] or mesh smoothing [37], or by assigning mixed properties to voxels which occupy an interphase region consisting of both yarn and matrix [38] and [39].

5. CONCLUSIONS

Three different geometrical models of a complex orthogonal 3D woven composite have been assessed using the voxel method and a continuum damage model for determination of elastic moduli and final failure. The assumptions used in an idealised model have been shown to produce significantly non-conservative results for stiffness and especially strength predictions. Models considering realistic geometry from simulation at two levels of compaction produced reduced, and more accurate, predictions. The fully compacted simulated geometry model produced reduced properties compared to the normalised lower compaction simulated geometry model due to the increase in waviness levels. The significant variation of results between the different models clearly illustrates the importance of textile geometry and that idealisation assumptions are not adequate for textiles with complex internal architecture. The fully compacted model should have produced closest match with experimental data, though results were slightly conservative. This has been attributed to the voxel discretisation of the geometry, as the stepped yarn surfaces in the simulated geometry models caused a reduction in stiffness and some fluctuation in strength. Considering such complex geometry introduces greater requirements on mesh refinement in order to achieve convergence than it was feasible to achieve for the considered unit cell and alternative strategies other than mesh refinement need to be considered to mitigate its effects.

ACKNOWLEDGMENTS

The authors wish to acknowledge the support of the Engineering and Physical Sciences Research Council (EPSRC) as well as Rolls-Royce plc through the Composites University Technology Centre (UTC) of the University of Bristol. Support of the University of Nottingham and permission from BAE Systems to use experimental data from reference [34], which was generated under the TSB funded 3D SIMCOMS project, is also gratefully acknowledged.

REFERENCES

1. Sherburn M. Geometric and mechanical modelling of textiles [PhD thesis]: University of Nottingham; 2007.

2. Lin H, Zeng X, Sherburn M, Long AC, Clifford MJ. Automated geometric modelling of textile structures. Text Res J 2012;82:1689–702.

3. Vandeurzen P, Ivens J, Verpoest I. A three-dimensional micromechanical analysis of woven-fabric composites: I. Geometric analysis. Compos Sci Technol 1996;56:1303–15.

4. Verpoest I, Lomov SV. Virtual textile composites software WiseTex: integration with micro-mechanical, permeability and structural analysis. Compos Sci Technol 2005;65:2563–74.

5. Cox BN, Dadkhah MS, Morris WL. On the tensile failure of 3D woven composites. Compos Part A: Appl Sci Manuf 1996;27:447–58.

6. Mahadik Y, Hallett SR. Effect of fabric compaction and yarn waviness on 3D woven composite compressive properties. Compos Part A: Appl Sci Manuf 2011;42:1592–600.

7. Green SD, Long AC, El Said BSF, Hallett SR. Numerical modelling of 3D woven preform deformations. Compos Struct 2014;108:747–56.

8. Zhou GM, Sun XJ, Wang YQ. Multi-chain digital element analysis in textile mechanics. Compos Sci Technol 2004;64:239–44.

9. Crookston JJ, Kari S, Warrior NA, Jones IA, Long AC. 3D textile composite mechanical properties prediction using automated FEA of the unit cell. In: ICCM-16. Kyoto; July 2007.

10. Jiang W-G, Hallett S, Wisnom M. Development of domain superposition technique for the modelling of woven fabric composites. Mechanical response of composites. Netherlands: Springer; 2008. p. 281–91.

11. Tabatabaei SA, Lomov SV, Verpoest I. Assessment of embedded element technique in meso-FE modelling of fibre reinforced composites. Compos Struct 2014;107:436–46.

12. Iarve EV, Mollenhauer DH, Zhou EG, Breitzman T, Whitney TJ. Independent mesh method-based prediction of local and volume

average fields in textile composites. Compos Part A: Appl Sci Manuf 2009;40:1880–90.

13. Belytschko T, Parimi C, Moës N, Sukumar N, Usui S. Structured extended finite element methods for solids defined by implicit surfaces. Int J Numer Methods Eng 2003;56:609–35.

14. Kästner M, Müller S, Ulbricht V. XFEM modelling of inelastic material behaviour and interface failure in textile-reinforced composites. Procedia Mater Sci 2013;2:43–51.

15. Kim HJ, Swan CC. Voxel-based meshing and unit-cell analysis of textile composites. Int J Numer Methods Eng 2003;56:977–1006.

16. Matveev MY, Long AC, Jones A. Mechanical properties of textile composites with variability in micromechanical properties. In: 15th European conference on composite materials. Venice; 2012.

17. Hallett SR, Biragoni P. Damage prediction for 3D woven composite structural features. In: TEXCOMP-10. Lille; 2010. p. 36–43.

18. Green SD, Long AC, Hallett SR. Simulated textile geometry with realistic deformations for finite element modelling. In: TexComp-11. Leuven; September 2013.

19. Sklansky J. Finding the convex hull of a simple polygon. Pattern Recogn Lett 1982;1:79–83.

20. Avis D, Bremner D, Seidel R. How good are convex hull algorithms? Comput Geom 1997;7:265–301.

21. Römelt P, Cunningham PR. A multi-scale finite element approach for modelling damage progression in woven composite structures. Compos Struct 2012;94:977–86.

22. Li S, Warrior N, Zou Z, Almaskari F. A unit cell for FE analysis of materials with the microstructure of a staggered pattern. Compos Part A: Appl Sci Manuf 2011;42:801–11.

23. Searles K, Odegard G, Kumosa M. Micro- and mesomechanics of 8-harness satin woven fabric composites: I – evaluation of elastic behavior. Compos Part A: Appl Sci Manuf 2001;32:1627–55.

24. Karahan M, Lomov SV, Bogdanovich AE, Mungalov D, Verpoest I. Internal geometry evaluation of non-crimp 3D orthogonal woven carbon fabric composite. Compos Part A: Appl Sci Manuf 2010;41:1301–11.

25. Ivanov DS, Lomov SV. Modeling of 2D and 3D woven composites. In: Polymer composites in the aerospace industry. Woodhead Publishing Limited; 2013.

26. Green SD, Matveev MY, Long AC, Hallett SR. Modelling of 3D woven composites with realistic unit cell geometry. In: ICCM-19. Montreal; July, 2013.

27. Asp L, Berglund L, Gudmundson A. Effects of a composite-like stress state on the fracture of epoxies. Compos Sci Technol 1995;53:27–37.

28. Fiedler B, Hojo M, Ochiai S, Schulte K, Ando M. Failure behavior of an epoxy matrix under different kinds of static loading. Compos Sci Technol 2001;61:1615–24.

29. Soden PD, Hinton MJ, Kaddour AS. Lamina properties, lay-up configurations and loading conditions for a range of fibre-reinforced composite laminates. Compos Sci Technol 1998;58:1011–22.

30. Djordjevic´ IM, Sekulic´ DR, Mitric´ MN, Stevanovic´ MM. Non-Hookean elastic behavior and crystallite orientation in carbon fibers. J Compos Mater 2010;44:1717–27.

31. Bogdanovich AE, Karahan M, Lomov SV, Verpoest I. Quasi-static tensile behavior and damage of carbon/epoxy composite reinforced with 3D noncrimp orthogonal woven fabric. Mech Mater 2013;62:14–31.

32. Ruijter W. Analysis of mechanical properties of woven textile composite as functions of textile geometry, University of Nottingham; 2008.

33. Caddell RM, Raghava RS, Atkins AG. Pressure dependent yield criteria for polymers. Mater Sci Eng 1974;13:113–20.

34. Townsend D, Jenkins S, Rezai A, Fishpool D, Siggs E, Middleton A. 3DSIMCOMS dissemination report BAE Systems; 2009.

35. Miyagawa H, Mase T, Sato C, Drown E, Drzal LT, Ikegami K. Comparison of experimental and theoretical transverse elastic modulus of carbon fibers. Carbon 2006;44:2002–8.

36. Maurin R, Davies P, Baral N, Baley C. Transverse properties of carbon fibres by nano-indentation and micro-mechanics. Appl Compos Mater 2008;15:61–73.

37. Boyd SK, Müller R. Smooth surface meshing for automated finite element model generation from 3D image data. J Biomech 2006;39:1287–95. [38] Zeng T, Wu L-Z, Guo L-C. Mechanical analysis of 3D braided composites: a finite element model. Compos Struct 2004;64:399–404.

38. Crookston JJ, Ruijter W, Long AC, Jones IA. Modelling mechanical performance including damage development for textile composites using a grid-based finite element method with adaptive mesh refinement. In: TEXCOMP-8. Nottingham; 2006.

39. Toho Tenax America Inc. Continuous filament carbon fiber data sheet. [cited 19th July 2013].

40. Chamis CC. Simplified composite micromechanics equations for strength, fracture toughness, impact resistance and environmental effects. DTIC Document; 1984.

41. Tsai CL, Daniel IM. Determination of shear modulus of single fibers. Exp Mech 1999;39:284–6.

42. Villeneuve JF, Naslain R. Shear moduli of carbon, Si–C–O, Si–C–Ti–O and alumina single ceramic fibers as assessed by torsion pendulum tests. Compos Sci Technol 1993;49:191–203.

43. Umeco. MVR444 resin data sheet. [cited 19th July 2013].

Index

W